Ecological Studies, Vol. 107

Analysis and Synthesis

Edited by

O.L. Lange, Würzburg, FRG
H.A. Mooney, Stanford, USA
H. Remmert, Marburg, FRG

Ecological Studies

Volumes published since 1989 are listed at the end of this book.

José M. Moreno   Walter C. Oechel
Editors

The Role of Fire
in Mediterranean-Type
Ecosystems

With 65 illustrations including 3 in color

Springer-Verlag
New York Berlin Heidelberg London Paris
Tokyo Hong Kong Barcelona Budapest

José M. Moreno
Departamento de Ecología
Facultad de Biología
Universidad Complutense
28040 Madrid
Spain

Walter C. Oechel
Department of Biology
San Diego State University
San Diego, CA 92182
USA

Cover illustration. Map of documented fires occurring between 1938 and 1985 in Burton Mesa, California. Adapted from Figure 7-4, page 128.

Library of Congress Cataloging-in-Publication Data
The role of fire in Mediterranean-type ecosystems / [edited by] José
  M. Moreno, Walter C. Oechel.
    p.   cm. – (Ecological studies; v. 107)
    Includes bibliographical references and index.
    ISBN-13:978-1-4613-8397-0
    1. Fire ecology.   2. Mediterranean climate.   3. Plants, Effect of
fires on.   I. Moreno, José Manuel.   II. Oechel, W. C.   III. Series.
QH545.F5R65   1994
574.5' 222 – dc20   93-42622.

Printed on acid-free paper.

Production coordinated by Chernow Editorial Services, Inc., and managed by Ellen Seham; manufacturing supervised by Genieve Shaw.
Typeset by Best-set Typesetter Ltd., Hong Kong.

9 8 7 6 5 4 3 2 1

ISBN-13:978-1-4613-8397-0   e-ISBN-13:978-1-4613-8395-6
DOI: 10.1007/978-1-4613-8395-6

Preface

During the last decades, the number of wildfires and the surface burned has increased in most countries of the Mediterranean region. Burned landscapes have become a prominent feature of many areas, and forest fires have become a major concern for politicians and the general public. Although fires have been present in the Mediterranean region for millennia, it is likely that changes in land use brought about by industrialization and/or other socioeconomic changes may have affected when, where, how big, how frequent, and how intense these fires are. Similar changes in pattern, extent, and intensity of fire may have also occured in other regions with Mediterranean-type climate. Until now, in most countries fire suppression is the main policy and large and increasing sums are spent every year to fight fires. To introduce new policies that would reduce fire risk and/or its impacts, it is necessary to have a good understanding of how fire affects the structure and functioning of ecosystems. The purpose of this book is to evaluate the main ways in which fire affects ecosystem properties in Mediterranean-type areas, in particular those from the Northern Hemisphere, and what ecological principles management policies should take into account to reduce fire related hazards.

A rich variety of vegetation types covers the landscapes of the Mediterranean region. L. Trabaud (Montpellier, France) reviews the postfire vegetation dynamics of different community types from studies done across the Mediterranean basin. The conclusion that emerges is that plant recovery is rapid and that the prefire community is restored mainly from

v

endogenous sources. Thus, it seems that most vegetation types are well adapted to overcome the effects of fire.

Unlike in the other Mediterranean-type regions of the world, in Chile there is no history of fires associated with lightning. Nevertheless, human-made fires are common today. Because of this lack of history, the comparison of the responses to fire of the Chilean matorral with those of other Mediterranean-type areas with a history of lightning fires (E. Fuentes, A.M. Segura, and M. Holmgren, Santiago, Chile) could help us understand the role of fire in the selection of plant adaptive traits.

Among the various components of the fire regime, the relationship between fire intensity (i.e., the total energy released during burning), and plant response is difficult to study either because determining the intensity of a wildfire at a specific place is not easy or because modifying it in experimental fires at scales larger than what can be done with a burner is complicated. J.M. Moreno and W.C. Oechel (Madrid, Spain, and San Diego, CA, USA) present the results of a manipulative experiment in which fuel load of a southern Californian chaparral was changed to produce different fire intensities. They show that increased fire intensity can have profound direct and indirect effects on the responses of the plant to fire.

The information available on postfire animal dynamics from the various Mediterranean-type areas of the world is still rather limited. This is reviewed by R. Quinn (Pomona, CA, USA), in particular for small mammals, showing that the pattern is not similar for all Mediterranean-type areas. Additionally, he presents results on the role of animals on plant development from experiments conducted in southern California. He shows that herbivory, as exemplified in the case of *Ceanothus crassifolius*, exerts a powerful influence on the postfire plant recovery process.

Fire alters the physics, chemistry, and biology of soil and directly or indirectly contributes greatly to modify the nutrient status of the soil. The relative impact of fire on soils in Mediterranean-type areas is magnified by the fact that fires occur in steep and poor soils. N.L. Christensen (Durham, NC, USA) reviews how fires have an impact on the soil, and points out some of the gaps we have in this area.

Precipitation is limited in Mediterranean-type areas. Hence, water is a limiting resource, not only for plants but also for people. Understanding how fire affects hydrology (S. Rambal, Montpellier, France) is critical in these areas since fires affect an increasingly large surface every year. This is even more compelling in light of impending climate change, as the combination of increased temperature and reduced precipitation may make water more limiting, which may further increase fire risk and, therefore, the surface affected by it.

Fires flourish where specific types and spatial arrangements of vegetation (i.e., fuel structure) occur. Fire, in turn, affects both the structure and spatial arrangement of the vegetation. Although the relationship between fire and vegetation structure is better known, we know little about the interactions between fire and the structure of the landscape. F.W. Davis and D.A. Burrows (Santa Barbara, CA, USA) provide a modeling tool to explore these relationships. They show that their model can help in understanding the dynamics of landscapes affected by fire and under strong human pressure.

Although in many Mediterranean-type areas there are a large number of fires, just a small percentage accounts for a very large fraction of the total area burned, and is responsible for the greatest losses in natural and human values. Designing a management strategy to reduce the occurrence of such catastrophic fires is a priority. To do so, we need to know in detail the impacts associated to various types of fires. P.J. Riggan and colleagues (Riverside, CA, USA) use the extensive knowledge from the San Dimas experimental forest to evaluate the risks associated with fires occurring at different stages of development of the chaparral and under different conditions. Their analysis provides the basis for a management policy that should include the use of fire to reduce fire hazard.

Understanding the role of fire in the Mediterranean region and how it can affect the conservation of its fauna, flora, and landscapes is critical, given the current trend in land use and increasing losses due to wildfires. Z. Naveh (Haifa, Israel) discusses the beneficial and detrimental effects of fire. He provides the elements for designing a management policy directed at the conservation of our environment.

This book was inspired by a postgraduate course held in Valencia (Spain) under the auspices of Universidad Internacional Menéndez Pelayo. We greatly appreciate the support of UIMP and its promotion of a friendly and efficient environment conducive to productive discussion and information exchange among the authors, students, and other interested parties.

José M. Moreno
Walter C. Oechel

Contents

Contributors

James A. Brass

NASA Ames Research Center
Ecosystem Science and Technology
    Branch
Moffett Field, CA 94035, USA

Fred E. Brooks

Department of Plant Pathology
University of California
Riverside, CA 92521, USA

David A. Burrows

Environmental Systems Research
    Institute
380 New York Street
Redlands, CA 92373, USA

Norman L. Christensen

School of the Environment
Duke University
Durham, NC 27708, USA

Frank W. Davis

Department of Geography
University of California
Santa Barbara, CA 93106, USA

Scott E. Franklin   25059 High Spring Avenue
                                       Santa Clarita, CA 91321, USA

Eduardo R. Fuentes   Departamento de Ecología
                                       P. Universidad Católica de Chile
                                       Casilla 114-D
                                       Santiago
                                       Chile

                                       Current address:
                                       United Nations Development
                                           Programs
                                       One United Nations Plaza
                                       New York, NY 10017, USA

Milena Holmgren   Departamento de Ecología
                                       P. Universidad Católica de Chile
                                       Casilla 114-D
                                       Santiago
                                       Chile

José M. Moreno   Departamento de Ecología
                                       Facultad de Biología
                                       Universidad Complutense
                                       28040 Madrid
                                       Spain

Zev Naveh   Faculty of Agriculture and
                                           Engineering
                                       Technion, Israel Institute of
                                           Technology
                                       32000 Haifa
                                       Israel

Walter C. Oechel   Department of Biology
                                       San Diego State University
                                       San Diego, CA 92182, USA

Ronald D. Quinn   Department of Biological Sciences
                                       California State Polytechnic
                                           University
                                       Pomona, CA 91768, USA

Serge Rambal

Centre d'Ecologie Fonctionnelle
   et Evolutive
Centre Emberger CNRS
Route de Mende
B.P. 5051
34033 Montpellier
France

Philip J. Riggan

USDA Forest Service
Pacific Southwest Research Station
Forest Fire Laboratory
4955 Canyon Crest Drive
Riverside, CA 92507, USA

Alejandro M. Segura

Departamento de Ecología
P. Universidad Católica de Chile
Casilla 114-D
Santiago
Chile

Louis Trabaud

Centre d'Ecologie Fonctionnelle
   et Evolutive
Centre Emberger CNRS
Route de Mende
B.P. 5051
34033 Montpellier
France

1. Postfire Plant Community Dynamics in the Mediterranean Basin

Louis Trabaud

Fire is such an ancient, universal, and ecological force that it has played an important part in shaping many of the vegetative communities and much of the landscape in the Mediterranean basin. The aridity of the climate has undeniably influenced, but not always preponderantly, the structure and composition of vegetations. Drought, combined with the nutrient poorness of the various soils, and fire contributed to the creation of such communities as matorrales and tomillares in Spain, maquis and garrigues in southern France, macchia in Italy, and xerovuni and phryganas in Greece. These various landscapes have been also modeled by human action, which has had a strong influence in these areas.

Just how ancient fire is cannot be known for certain, but there can be no doubt that its first occurrence predates humans by a considerable margin; it may be as old as the terrestrial vegetation (Harris 1958; Komarek 1973). Lightning is one of the natural causes of vegetation fires, from the tundra to tropical forests (Komarek 1964, 1967, 1968; Taylor 1969). In the same way, other nonanthropogenic phenomena, particularly volcanic eruptions, may have also caused fires.

At first, fire was a natural component that appeared more or less regularly in the natural cycle of vegetation succession. Its advent permitted the rejuvenation of some stands and created a mosaic of plant communities. However, the appearance of humans disrupted this balance of nature, substituting an artificial situation and upsetting the previous order. Humans have used and misused fire, and so fire, when combined

1

with forest felling, grazing by domestic animals, and extensive but agressive agriculture (uprooting of plants for cultivation), has contributed to shaping the vegetational landscapes of today.

In many Mediterranean ecosystems, fire controls age, structure, and species composition. Fire acts with different frequencies and intensities, depending on the vegetation and climatic situations. Thus, vegetation composition and structure depend on climate, fire frequency, and intensity, while fire frequency and intensity, in turn, rely on vegetation structure and climatic regimes.

Old Concepts

Research dealing with the problem of wildland fires is characterized by two periods: a pioneer one during which there were very few works concerning this subject, and a more recent period where the problem of fire was tackled with rigorous methods and techniques.

The pioneer period can also be divided into two parts: a first one during which publications discussed fire without thoroughly studying its effect on the ecosystems, and a second one, during which most scientists, under the influence of Clementsian and phytosociological ideas, were anxious at all costs to determine that the vegetation dynamics after fire followed a succession of stages or "associations" in a single linear process, without paying attention to the responses of plants constituting the communities.

The first period was characterized by the works of precursors. Ribbe (1865, 1869) recorded the importance of the understory in the propagation of fire; he documented the dominance of *Erica arborea* and the rapid colonization of burned sites by *Cistus* spp. He also recorded the action of pine cones in propagating wildfires, and particularly discussed the causes of wildfires and the jurisdiction problems linked with them.

After Ribbe, other authors discussed fire and drew the same conclusions about its influence. Jacquemet (1907) attributed deforestation to wildfires; Cotte (1911) discussed the effect of fire on *Quercus suber* and *Pinus pinaster*. Flahault (1924) studied the phenomenon more thoroughly, citing the effects of fire on vegetation and soil; he recorded the first species establishing in burned areas – *Cytisus purgans*, *Erica arborea*, *Cistus monspeliensis*, and *Quercus coccifera*.

Ducamp (1932) and Laurent (1937) tried to explain the effect of wildfires on vegetation and more particularly considering their impact on erosion and forest "degradation." Except for Ribbe, the forerunner, practically all of these authors made few precise observations or careful studies on the wildland fire phenomenon and the havoc wreaked on vegetation. These authors' results were only general and fragmentary comments.

In the 1930s, the setting up of the phytosociological method gave rise to a recrudescence of activity for naturalists, who by then possessed a tool allowing them to study the fire impact on vegetation more precisely. Associated with the North American Clementsian theory, these researchers considered succession as an unidirectional change of vegetation types, each successive type establishing itself because the preceding type had modified the site in a way favorable to its successor; this sequence naturally ended in a climax type that was stable and self-maintaining under the current conditions of site and climate. This was seen as a deterministic process. If the climax was destroyed by a catastrophic event (e.g., fire), community recovery had to go through the sequence of all the preceding stages before the same climax could be re-established, provided the climate was unchanged.

During this latter period, the phytosociological school dominated in Europe. Braun-Blanquet (1935, 1936) looked upon wildfire as being as important a factor as tree felling and overgrazing in the regressive dynamics of vegetation. It was his opinion that the degradation process of the *Quercus ilex* forest, by successive associations from the climax forested to grass swards, considered as the ultimate stage in the regression series.

By comparing different types of plant associations, Kornas (1958) studied the causes of the regressive succession in the Gardiole mountain near Montpellier, France. He described the same stages as Braun-Blanquet. In Provence, Molinier (1953, 1968) and Molinier and Molinier (1971) also considered this problem, but they did not try to analyze the process of vegetative dynamics after fire. They were satisfied with looking at wildfire as a factor contributing to the degradation of plant communities and performed studies similar to those of Kornas (1958).

During that period, one author particularly stands out. Kuhnholtz-Lordat (1938, 1958) did not accept the strict phytosociological theory, but rather viewed fire as a tool of human hands that changed wildlands into agricultural fields. Kuhnholtz-Lordat (1938) was the first to define the term pyrophytes, and to provide a list of pyrophytic communities. He focused on the resistance of species to fire and their ability to regenerate after fire. Kuhnholtz-Lordat (1952) claimed that the vegetation of Provence was almost totally of pyrophytic origin, and he cited four primary pyrophytes: *Quercus coccifera*, *Erica multiflora*, *Cistus monspeliensis*, and *Pinus halepensis*. He associated soil degradation with fire, and confirmed that the persistence of vegetation in the Mediterranean area depends on frequent fires – short periodicity where only vegetatively regenerating species can survive; and long periodicity where obligate seeders can reproduce.

Later, Kuhnholtz-Lordat (1958) defined more precisely the term and classification of pyrophytes: (1) pyrophytes with passive resistance, because of their constitution and high water content in their tissues (succulents, e.g., *Agave*) or thick bark (*Quercus suber*); (2) vegetatively regenerating pyrophytes with the capability to sprout after destruction of the above-ground organs by epicormic shoots (as *Arbutus unedo*) or by below-ground resprouts (*Quercus coccifera*, *Pteridium aquilinum* and several perennial grasses); (3) pyrophytes with an indirect resistance that creates an unfavorable environment for plants around them; and (4) social pyrophytes that regenerate by seed (e.g., Mediterranean *Pinus* spp., *Cistus* spp.). However, he still gave flow diagrams of succession without studying and providing evidences of the real process of vegetation dynamics after fire. Barry (1960), following identical concepts, joined his study near Nîmes and presented diagrams of several successive stages.

All the preceding authors considered that fire created a series of increasingly degraded successive stages. They only compared, by synchronic way different associations thought a priori to follow a succession. The actual processes by which vegetation recovers after fire were not precisely tackled. Even today, some authors present the same flow diagrams without providing any precise data from field observations, starting from "original" forests down to shrublands and open swards (Le Houérou 1974; Macchia et al. 1984).

Recent Research and New Results

At present, several detailed researchers have started working in the European part of the Mediterranean basin and the results give new perspectives on plant community dynamics after fire.

Spain

In southwest Andalusia, Garcia Novo (1977) described the successional stages after fire of one type of matorral growing in the Doñana National Park. He identified five stages. During the first stage (0 to 3 mo) no seed germination, no annuals appeared; only a few resistant species recovered, with two proeminent plants: *Chamaerops humilis*, and *Daphne gnidium*. In the second stage (first year), both woody and herb species germinated. Fire-resistant species started their growth. Many annuals were present. Some woody perennials started growth from stumps. Nonsprouting species (obligate seeders) profusely germinated at that time, producing many young plantlets. During the third stage (second year), all the species grew steadily; no germination of seeds from woody species was observed. Grasses reached their maximum development. During the fourth stage (third to fourth years) the shrubs increased dramatically; the matorral structure evolved towards the mature matorral type; the importance of the herb layer declined. From the fifth year onwards (fifth stage) the matorral recovered its original vegetational composition and structure.

Northward, in the area of Alicante, Mansanet (1982) synchronically studied the development on 10 years of matorrales with *Rosmarinus officinalis* and *Ulex parviflorus* dominated by a tree layer of *Pinus hale-pensis* comparing different burned sites located on calcareous soils. He found the community was restored at the end of 10 years; only the Aleppo pine seedlings present in all the sites did not reach their adult size. Floristic richness did not seem to change through time. When he compared the life-form ratios during the years of the study, he stated that the proportions remained constant, except that a greater abundance of therophytes during the first 5 years showed an increase of annuals at that time. The average height of the community increased from 15 cm in the first year to 85 cm 10 years after fire.

In the region of Valencia, Sanroque et al. (1985) studied a dense matorral where *Erica multiflora*, *Rosmarinus officinalis* and *Ulex parviflorus* were the dominant species, overspread by a tree overstory of *Pinus halepensis* (thus a community similar to the one studied by Mansanet) from 1 to 2 years after fire. The number of observed species was higher in the burned area in comparison with the control site because of the presence of perennial and annual herbaceous species. The species reappearing in the burned area were the same as those present in the control.

In the mountain of Garraf (South of Barcelona) Papió (1984, 1985), observed that the rapid recovery on compact limestone of vegetation after fire was through sprouts or seedlings. Most of the species had appeared during the first 18 months. He also gave a good description of the phenological stages of the *Quercus coccifera* garrigue. The average height of this type of garrigue (a low shrubland) reached 50 cm in 1 year and 1 m at 8 years. The phytomass of the above-ground parts was $2.8 \, t.ha^{-1}$ at 1 year after fire and $16.5 \, t.ha^{-1}$ at 8 years. The regeneration of *Pinus halepensis* was also considered as well as the mortality of young seedlings.

In Majorca (Balearic Islands), Morey and Trabaud (1988) studied a shrub community in calcareous soil during the first 3 years. The tendency here was to reestablish the previous ecosystem. Practically all of the perennials resprouted; only 7% showed a seed dispersal strategy. Floristic composition was maximum during the first years after fire.

In *Erica arborea* and *Calycotome spinosa* maquis of Cape de Creus (northeast Catalonia), Franquesa (1987) reached similar conclusions: rapid recolonisation was basically from sprouts of prolific species, with a rapid stabilization of the vegetation state; the species that were dominant appeared during the first years of recolonisation and had been present before fire; exogenous invading plants (most often annuals) appeared and disappeared very quickly, remaining only for a couple of years during the first recovery. There was no actual succession process.

Farther west, in the province of Leon, in the understory of *Quercus pyrenaica* forests, specific diversity increased the first year mainly due to an increase in species richness. All the studied formations kept their own specific characteristics, constituting an autosuccession process (Tarrega and Luis-Calabuig 1987).

In a not truly Mediterranean area (Galicia), Casal (1985) and Casal et al. (1986) found that the recovery in matorrals dominated by *Ulex europaeus* was the most important from sprouts. A few woody species became dominant in the community because of this recolonization system; other species that recolonized by seeds (e.g., *Cistus* spp.) became less dominant after fire. The same occurred for herbs and grasses. Through time, the horizontal and vertical structures changed strongly: shrub height, cover, and phytomass increased as matorral was aging. Annual herbs were rapidly replaced by perennial herbs.

France

The dynamics of vegetation after fire was studied in the calcareous Garrigues zone of Bas-Languedoc by Trabaud (1970, 1980, 1982) and Trabaud and Lepart (1980, 1981). The vegetation of the burned communities returns quickly towards its initial state. Most often the species that were present 12 years after fire were the first ones to reappear and to become more and more numerous over time. Forty-seven plots were studied, located in eight types of plant communities, from forests to swards, representative of the vegetation of the area. The dynamics were studied by the diachronic method for 10 or 12 years.

One year after fire, 70% of the plots studied possessed more than 75% of the species that were present 10 or 12 years later. Two years after fire, this percentage was over 80%, and in 5 years it reached 100% (Figure 1–1). The reversion toward a metastable state, at least for the time considered, was quickly accomplished.

The study of the development of floristic richness subsequent to fire shows that the different communities followed a highly general model (Figure 1–2). During the first months after fire there were relatively few species. Floristic richness reached a maximum between the first and the third year after fire, and tended to finally stabilize after the third year.

The richness of the intermediate stages resulted from the presence of short-lived species that were progressively eliminated during vegetation recovery. Most often, these species are therophytes that normally did not belong to the communities. The generally higher number of species during the first 3 years could be attributed to the fire-induced opening of vegetation cover, the disappearance of the litter, and the richness in nutrients in the upper soil layer. These conditions favored the establishment of several alien species that later disappeared as plant cover closed up.

Most of the perennial plants of the communities studied were able to regenerate by sprouts; several can regenerate both by seed and sprout. Only a few of the woody perennials reproduce solely by seed, such as *Cistus monspeliensis*, *Pinus halepensis*, and *Rosmarinus officinalis*. Thus, some of them only appear in the spring following a fire. Species that can

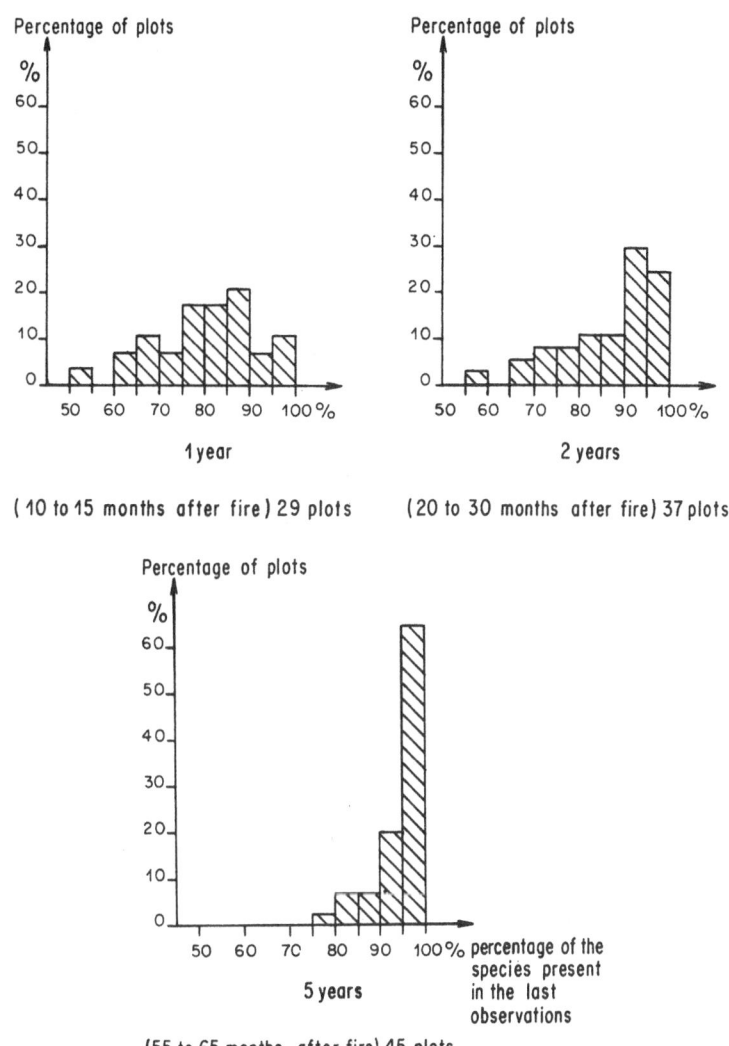

Figure 1–1. Three measurements of the relative importance of species present in the terminal communities. (From Trabaud 1980.)

regenerate vegetatively rapidly occupy the burned areas, and due to their competitive ability, prevent alien pioneer species from invading the areas. This dominance is typical for most of the studied communities (Table 1–1) since, except for the pine forest and the rosemary garrigue, practically 80% or more of the recorded species can immediately regenerate by sprouting. It is easier to understand why plant communities of the Mediterranean region reconstitute similarly to those that existed before fire.

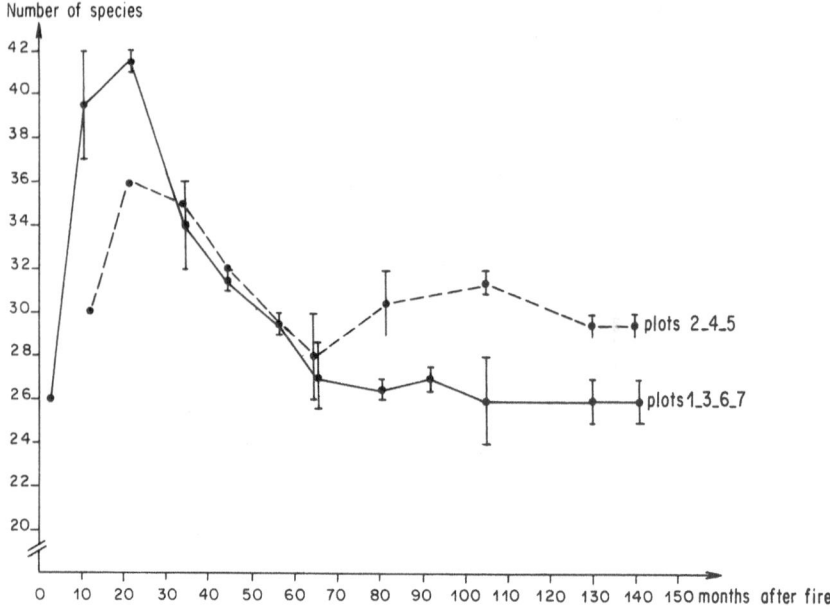

Figure 1–2. Dynamics of the floristic richness of dense *Quercus ilex* coppice-woods after fire. (From Trabaud 1980.)

Table 1–1. Percentage of the Hits Made by Species Preferentially or Exclusively Regenerating by Sprouts to the Total Vegetation Hits During Recovery after Wildfire

Species	Range of Hits in Percent of Total Vegetation	Percentage of Cases for the Cited Ranges
Quercus ilex dense woodlands	75–99	79.5 for the ranges 80 to 100%
Quercus ilex open woodlands	51–80	71.0 for the ranges 60 to 80%
Quercus coccifera dense garrigues	70–96	77.9 for the ranges 80 to 100%
Quercus coccifera open garrigues	62–97	66.7 for the ranges 80 to 100% ·
Pinus halepensis forests	31–79	52.2 for the ranges 40 to 60%
Rosmarinus officinalis garrigues	21–90	64.7 for the ranges 70 to 90%
Brachypodium ramosum swards	63–90	94.4 for the ranges 70 to 90%
Brachypodium phoenicoides swards	64–86	76.5 for the ranges 70 to 80%

This fact is important for a precise understanding of the vegetation recovery process and the vegetative landscape dynamics.

Trabaud (1980, 1982, 1983), studying the community structure, found after fire and through succeeding years that stand strata become more and

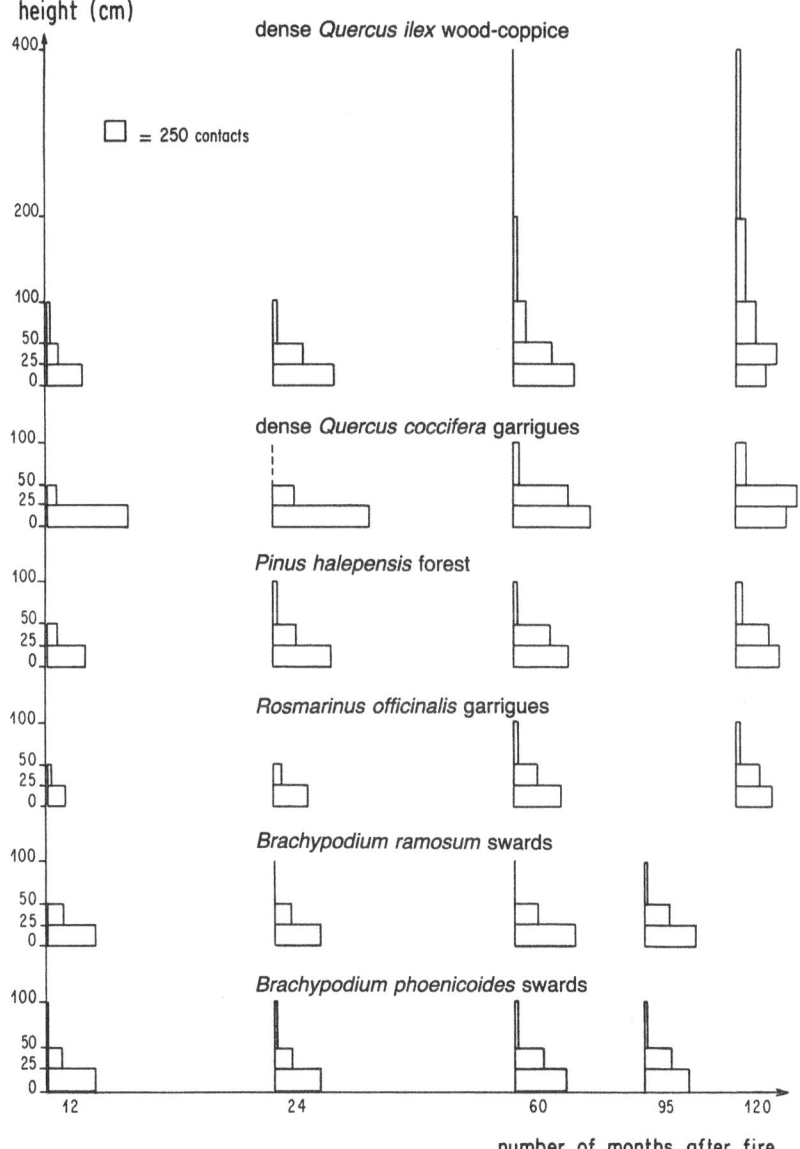

Figure 1–3. Stratification changes through time in some burned plant communities. (From Trabaud 1980.)

more complex and vegetation tends to grow up from the lower understory to the upper overstory (Figure 1–3). As communities progressively age, the importance of the lower layers decreases, whereas that of the upper layers increases. There was a progressive multiplicity of strata similar to those of origin, the higher strata appearing later in succession. During the 12 years of observation, *Quercus ilex* forest-coppices reached a fifth layer (from 2 to 4 m). *Pinus halepensis* forests (although belonging to forest communities) generally did not grow beyond the third layer (1 m). This difference is due to the type of survival trait used by the dominant species of each community to regenerate after fire. *Q. ilex* sprouts from stumps and grows rapidly, reaching heights of 2 m in 70 months, whereas *P. halepensis* can only reproduce by seed and thus shows a slower growth following fire (80 cm in 80 months).

The recovery of *P. halepensis* forests was more closely studied (Trabaud et al. 1985a, 1985b). Three phases of the increase of the understory phytomass could be distinguished: a first phase with a rapid increase lasting for 2 years; a second phase with a slower increase when only shrubs grew; and a third phase, with no increase, when the understory shrubs reached adult size. Thirty years after fire, understory phytomass ranged between 9 to 12 t.ha^{-1}. Litter predominantly woody during the first years after fire became foliar dominant as stands got older. Pine density increased to a maximum during the first 15 years (Figure 1–4). It then decreased, probably because of mortality brought about by in-

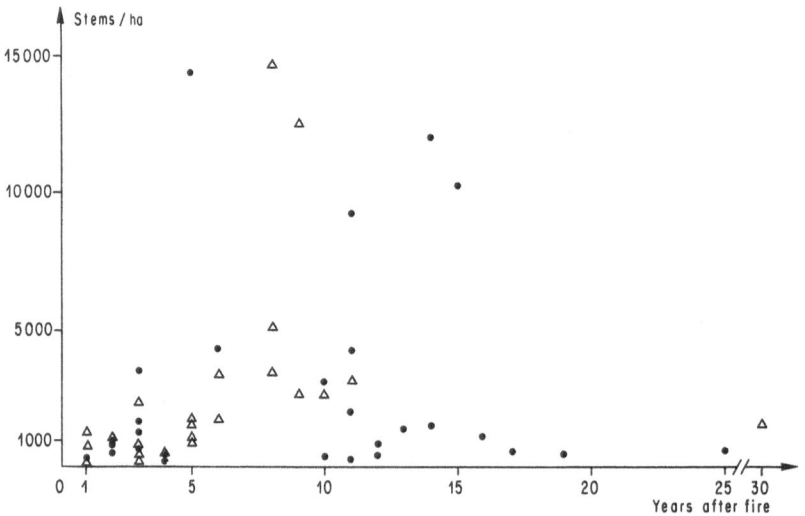

Figure 1–4. Changes of the number of pine trees through years after fire: △, *Quercus coccifera* or ●, *Rosmarinus officinalis* understories. (From Trabaud et al. 1985b.)

traspecific and interspecific competition. The areal distribution of young pines was uniform in all the sites studied, apparently due to the superimposition of seed rains from various sources.

In the siliceous mountains of the Albères and the Aspres (eastern French Pyrénées), Prodon et al. (1984), by studying six types of communities in a synchronic way, from grass swards to forests of *Q. ilex* or *Q. suber* in plots at 1 year to 4 years old after fire, found that reappearance of species was immediate. In the same time, previous species recovered; many were invasive during the first years after fire, including numerous annuals. Floristic richness was always higher in burned sites compared to unburned ones. These authors recorded that in the sward-to-forest community sequence, floristic richness was always lowest in forests. Moreover, during the first 2 years after fire, great modifications occurred in the proportion of life forms, with therophytes being extremely abundant. However, the dominant species that re-established after fire were species that were present before. Little-layered communities (i.e., swards, shrublands) rapidly reached a state comparable with the initial one.

Italy

In southern central Italy, most of the dominant and abundant plants of the macchia (characterized by *Pistacia lentiscus*, *Myrtus communis*, and *Ampelodesmos mauritanicus*) regenerated through vegetative organs. The perennial species belonging to the original community overtopped and eliminated alien invading species (De Lillis and Testi 1990; Mazzolini and Pizzolongo 1990). Plant diversity was highest during the second year after fire, when floristic richness was the highest. The vegetation's high resilience was due to the sprouting ability of the woody shrubs that were rapidly re-establishing.

Greece

Postfire regeneration of plant species in this part of the eastern Mediterranean basin has been studied in phryganas (Papanastasis 1977a, 1977b; Arianoutsou and Margaris 1981; Arianoutsou 1984). This type of vegetation is mostly characterized by spiny subshrubs such as *Sarcopoterium spinosum*, *Phlomis fruticosa*, *Euphorbia acanthothamnos*. Postfire regeneration occurred from below-ground organ resprouting and seed germination. The beginning of postfire succession started with herbaceous vegetation. Annual species dominated the phrygana burns in the first year and thereafter declined rather rapidly. Legumes (*Trifolium*, *Medicago*, etc.) were the prevailing annual species. Also quite abundant were annual grasses (Papanastasis 1977a, 1977b). These species can compete with heavily original phrygana seedlings appearing after fire. However, this situation gradually changed so that by the end of the seventh postfire year the relative structural importance of woody and herbaceous species was similar to what it was before.

The development of the structure of these low communities (less than 1-m high) was studied through the changes in phytomass. Papanastasis (1977b) found that production was $3 t.ha^{-1}$ in a *Phlomis fruticosa* community 8 months after fire, whereas in a *Sarcopoterium spinosum* phrygana, it was $1.2 t.ha^{-1}$ in the first year, $1.6 t.ha^{-1}$ the second year, and $2.8 t.ha^{-1}$ the third year. In a community with the same dominant species, Arianoutsou (1984) recorded a rapid growth during the first 5 years, up to $5 t.ha^{-1}$, after which the increase was slower. The proportion of the herb phytomass was greater during the first 2 years after fire, then declined rapidly, with shrubby species dominating from the third year onward.

Troumbis (1985), studying a period of 60 years by the synchronic method in *Cistus* stands (*C. villosus*, *C. salvifolius*), found floristic richness did not vary much along the succession stages; neither did diversity. On the contrary, there were great floristic interannual variations depending on whether two consecutive years were dry or humid.

The cover recovery of these *Cistus* low shrublands was rapid during the first years, then relatively stabilized after 10 years. *Cistus* cover was practically unchanged from the sixth year. Herbaceous species were very abundant during the first years, with a high cover; after 10 years this cover became and remained constant, its growth varying according to sites. The vegetative cover of woody species increased through time, but that of herbaceous plants varied greatly. During recovery, the complexity of layers became higher and higher. At the beginning of recolonization only lower layers (<40 cm) constituted the main part of vegetation; afterwards, higher layers (>40 cm) appeared and their importance increased more and more. In the same manner, foliar organs through time were located in higher vegetation levels, whereas supporting woody organs were more and more numerous in lower layers.

Conclusion

After fire, plants appear rapidly and cover the ground surface. Nearly all authors agree and reach the same conclusions:

(1) The abundance of herbaceous species (mostly annuals) is quite remarkable during the first years in the burned areas; (2) of the species that gain dominance during the re-establishment of the mature vegetation, the majority are present in the first few years after fire; (3) the establishment of previous communities is a rapid phenomenon; and (4) as burned communities age, returning to a state similar to that of unburned systems, structure becomes more and more complex, with numerous layers. Herbaceous layers predominant during the beginning stages decrease and are replaced by shrub and tree layers. In the same manner, phytomass increases from herbaceous to woody dominant.

In the most recent studies, authors agree that the development of vegetation after fire follows the "initial floristic composition" model described by Egler (1954) or the "inhibition" model from Connell and Slatyer (1977): All the prefire species are present immediately after fire, even if later on the relative abundance or frequency of individuals changes. Thus, there is no real succession, or floristic relays, or different communities on the same site, as is characteristic of secondary succession, but rather, an autosuccession process leading to a recovery of preburn communities. In fact, Mediterranean vegetation does present a recovery by "direct" endogenous process, (i.e., the species that existed before fire again occupy the burned sites, as opposed to an "indirect" or exogenous recovery characterizing a succession of stages as it was described for old field dynamics). The plants that persist are those that appear immediately after fire and that existed previously. Floristic composition and structure of burned communities tend to revert to a metastable equilibrium similar to that which existed without fire.

The plants of the Mediterranean region withstand fire by different survival, vegetative, and sexual traits. The present vegetation of the Mediterranean basin results from many years of evolution during which plants acquired mechanisms to overcome the effects of fire as well as climatic (especially summer drought) factors. This evolutionary impact has been shown by positive and negative feedback responses that enable direct fire tolerance or permit its avoidance, followed by vegetative and reproductive regeneration. Each type of species has evolved different survival traits allowing the plants to survive disturbances and perpetuate themselves as well as the communities of which they are a part.

References

Arianoutsou, M. 1984. Post-fire successional recovery of a phryganic (east Mediterranean) ecosystem. *Acta. Oecol. Plant.* 5:387–394.

Arianoutsou, M., Margaris, N.S. 1981. Early stages of regeneration after fire in a phryganic ecosystem (east Mediterranean). I. Regeneration by seed germination. *Biol. Ecol. Médit.* 8:119–128.

Barry, J.P. 1960. Contribution à l'étude de la végétation de la région de Nîmes. *Année Biol.* 36:311–550.

Braun-Blanquet, J. 1935. Un problème économique et forestier de la garrigue languedocienne. *Comm. Sigma.* 35:11–22.

Braun-Blanquet, J. 1936. La forêt d'Yeuse languedocienne (*Quercion ilicis*). Monographie phytosociologique. *Mem. Soc. Etud. Sci. Nat. Nimes* 5:147.

Casal, M. 1985. Cambios en la vegetación de matorral tras incendio, en Galicia. *In Estudios sobre Prevención y Efectos Ecológicos de los Incendios forestales.* Instituto Conservación Naturaleza, Madrid.

Casal, M., Basanta, M., Garcia Novo, F. 1986. Sucesión secundaria de la vegetación herbácea tras el incendio del matorral bajo repoblación forestal de *Pinus. Boletín Sociedad Española Historia Natural (Biología)* 82 1/4, 25–34.

Connell, J.H., Slatyer, R.O. 1977. Mechanisms of succession in natural communities and their role in community stability and organization. *Amer. Natur.* 111:1119–1144.

Cotte, J. 1911. Les forêts du massif archéen du Var. *Bull. Soc. Linnéenne Provence*, 140–141.

De Lillis, M., Testi, A. 1990. Post-fire dynamics in a disturbed Mediterranean community in central Italy. *In* M.G. Goldammer and M.T. Jenkins (eds.), *Fire in Ecosystem Dynamics*. SPB Acedemic Publishing. The Hague, 53–62.

Ducamp, R. 1932. Au pays des incendies. *Rev. Eaux Forêts* 70:380–393.

Egler, F.E. 1954. Vegetation science concepts. I. Initial floristic compostion, a factor in old field vegetation development. *Vegetatio* 4:412–417.

Flahault, C. 1924. Les incendies de forêts. *Rev. Bot. Apl. Agr. Colon.* 4:1–20.

Franquesa, T. 1987. Regeneracio de les brolles silicicoles de la peninsula del Cap de Creus. *Quadern Ecol. Apl.* 10:113–129. Barcelona.

Garcia Novo, F. 1977. The effects of fire on the vegetation of Doñana National Park, Spain. *Symp. Environm. Consequences Fire and Fuel Manage, Medit. Ecosyst.* USDA For. Serv. Gen. Tech. Rep. WO-3, 318–325.

Harris, T.M. 1958. Forest fire in the Mesozoic. *J. Ecol.* 46:447–453.

Jacquemet, C. 1907. Sur les incendies et le déboisement de nos collines. *Ann. Soc. Sci. Nat. Provence* 1:47.

Komarek, E.V. 1964. The natural history of lightning. *Annu. Tall Timbers Fire Ecol. Conf.* 3:139–183.

Komarek, E.V. 1967. The nature of lightning fires. *Annu. Tall Timbers Fire Ecol. Conf.* 7:5–41.

Komarek, E.V. 1968. Lightning and lightning fires as ecological forces. *Annu. Tall Timbers Fire Ecol. Conf.* 8:169–197.

Komarek, E.V. 1973. Ancient fires. *Annu. Tall Timbers Fire Ecol. Conf.* 12:219–240.

Kornas, J. 1958. Succession régressive de la végétation des garrigues sur les calcaires compacts dans la Montagne de la Gardiole, près de Montpellier. *Acta Soc. Bot. Pol.* 27:563–596.

Kuhnholtz-Lordat, G. 1938. *La terre incendiée*. Maison Carrée, Nîmes, France.

Kuhnholtz-Lordat, G. 1952. *Le tapis végétal dans ses rapports avec les phénomènes actuels de surface en Basse-Provence*. Le Chevallier, Paris, France.

Kuhnholtz-Lordat, G. 1958. L'écran vert. *Mém Mus. Natl. Hist. Nat.* 9:1–276.

Laurent, L. 1937. A propos des incendies de forêts en Basse-Provence. *Le Chêne* 44:139–148.

Le Houérou, H.N. 1974. Fire and vegetation in the Mediterranean basin. *Annu. Tall Timbers Fire Ecol. Conf.* 13:237–277.

Macchia, F., Lopinto, M., D'Amico, F.S. 1984. Fuoco e successione della vegetazione nei boschi puri e misti de *Pinus halepensis* Mill. del Gargano. *Convegno Int. Studi Probl. Incendi Boschivi in Ambiente Medit. Assess. Agr. Forest. Reg.* Puglia, Bari, 539–553.

Mansanet Terol, C.M. 1982. Contribución al estudio de la evolución de la vegetación tras el incendio forestal en algunas comarcas de la provincia de Alicante. Approximación a la problemática de los incendios forestales en esta provincia. *Mem. Licenciatura, Ciencias Biol.* Univ. Valencia.

Mazzoleni, S., Pizzolongo, P. 1990. Post-fire regeneration patterns of Mediterranean shrubs in the Campania region, Southern Italy, *In* J.G. Goldammer and M.J. Jenkins (eds.), *Fire in Ecosystem Dynamics*. SPB Academic Publishing. The Hague, 43–51.

Molinier, R. 1953. Le feu et l'avenir des forêts de Provence. *Rev. Gen. Sci. pures et appliquées* 60:199–208.

Molinier, R. 1968. La dynamique de la végétation provençale. *Collect. Bot.* 7:817–844.

Molinier, R., Molinier, R. 1971. La forêt méditerranéenne en Basse-Provence. *Bull. Mus. Hist. Nat. Marseille* 31:76.

Morey, M., Trabaud, L. 1988. Primeros resultados sobre la dinamica de la regetacion tras incendio en Mallorca. *Studia Oecologica*, 5:137–159.

Papanastasis, V.P. 1977a. Early succession after fire in a maquis-type brushland of northern Greece. *Forest* 30:19–26.

Papanastasis, V.P. 1977b. Fire ecology and management of phrygana communities in Greece. *Symp. Environm. Consequences Fire and Fuel Manage. in Medit. Ecosyst.* USDA Forest Serv. Gen. Tech. Rep. WO-3, 476–482.

Papió, C. 1984. *La regeneracio de la vegetacio despres de l'incendi de juliol de 1982 al Parc Natural de Garraf.* Dept. Ecol., Univ. Aut. Barcelona.

Papió, C. 1985. *La regeneracio de la vegetacio del massis de Garraf despres de l'incendi de juliol de 1982.* Dept. Ecol., Univ. Aut. Barcelona.

Prodon, R., Fons, R., Peter, A.M. 1984. L'impact du feu sur la végétation, les oiseaux et les micro-mammifères dans diverses formations méditerranéennes des Pyrénées Orientales: premiers résultats. *Rev. Ecol. Terre et Vie.* 39: 129–158.

Ribbe, C. 1865. Des incendies dans les forêts résineuses du département du Var. Rev. Agr. Forest. Provence, 1–7, 17–29, 89–115, 145–163, 181–196, 213–233.

Ribbe, C. 1869. Enquête sur les incendies de forêts dans les Maures et l'Esterel. Rev. Agr. Forest. Provence, 111–119.

Sanroque, P., Rubio, J.L., Mansanet, J. 1985. Efectos de los incendios forestales en las propiedades del suelo, en la composicion floristica y en la erosión hídrica de zonas forestales de Valencia (España). *Rev. Ecol. Biol. Sol.* 22:131–147.

Tárrega, R., Luis-Calabuig, E. 1987. Effects of fire on structure dynamics and regeneration of *Quercus pyrenaica* ecosystems. *Ecologia Mediterranea* 13: 79–86.

Trabaud, L. 1970. Quelques valeurs et observations sur la phytodynamique des surfaces incendiées dans le Bas-Languedoc (premiers résultats). *Natur. Monspel.* 21:231–242.

Trabaud, L. 1980. Impact biologique et écologique des feux de végétation sur l'organisation, la structure et l'évolution de la végétation des garrigues du Bas-Languedoc. Thèse Doct. Etat Univ. Sci. Tech. Languedoc Montpellier.

Trabaud, L. 1982. Effects of past and present fire on the vegetation of the French Mediterranean region. *Symp. Dynamics, Manage. Medit. Type Ecosyst.* USDA Forest Serv. Gen. Tech. Rep. PSW-58:450–457. Pacific Southwest Forest and Range Exp. Stn.

Trabaud, L. 1983. Evolution après incendie de la structure de quelques phytocénoses méditerranéennes du Bas-Languedoc (Sud France). *Ann. Sci. Forest.* 40:177–195.

Trabaud, L., Lepart, J. 1980. Diversity and stability in garrigue ecosystems after fire. *Vegetatio* 43:49–57.

Trabaud, L., Lepart, J. 1981. Floristic changes in a *Quercus coccifera* L. garrigue according to different fire regimes. *Vegetatio* 46:105–116.

Trabaud, L., Grosman, J., Walter, T. 1985a. Recovery of burned *Pinus halepensis* Mill. forests. Understorey and litter phytomass development after wildfire. *Forest Ecol. Manage.* 12:269–277.

Trabaud, L., Michels, C., Grosman, J. 1985b. Recovery of burned *Pinus halepensis* Mill. forests. Pine reconstitution after wildfire. *Forest Ecol. Manage.* 13:167–179.

Troumbis, A. 1985. Dynamique après perturbation des populations de deux espèces de *Cistus* à réproduction sexuée obligatoire. Thèse Doct. Univ. P. Sabatier, Toulouse.

2. Are the Responses of Matorral Shrubs Different from Those in an Ecosystem with a Reputed Fire History?

Eduardo R. Fuentes, Alejandro M. Segura, and Milena Holmgren

The California chaparral has become a model for how terrestrial vegetation adapted to natural fires can respond to this disturbance. The large amount of research done to date on the California chaparral and the diversity of its responses have transformed this ecosystem into a model, especially for the five Mediterranean-type communities. Of particular importance here has been the relatively high frequency of summer lightning as a source of natural ignition and thus the possibility of using the conceptual tools of evolutionary biology in the interpretation of the various responses.

In this chapter we examine the reponses of the Chilean matorral shrubs to fires. Unlike California, central Chile has no summer lightning, mainly because of the high altitude of the Andes ranges and the strength of the anticyclone of the southern Pacific, both of which prevent tropical storms from drifting toward the south during summer. This feature of the Chilean matorral is unique among the five areas with a Mediterranean-type climate. All others seem to have a current history of summer lightning as a source of natural (pre-human) fires. Today, fires in central Chile are human-made.

We will address the following two questions: (1) How do matorral ecosystems respond to fire? and (2) Are their responses any different from those of Mediterranean-type ecosystems with a recognized evolutionary history of fires, such as the California chaparral?

The Central Chile Scenario

In Chile, conditions for a Mediterranean-type climate (Aschmann 1973) extend from the vicinity of Combarbalá (approximately 31°S) to Linares (approximately 35°S) (Figure 2–1). As expected from general geographical considerations, its altitudinal limits decrease with latitude. At the latitude of Santiago (33°27'S) the Mediterranean-type climate prevails between sea level and about 1500 m; above this altitude, winter is too cold and precipitation is mostly in the form of snow.

Two mountain ranges running parallel to the Pacific coast are primary structuring factors of central Chile landscapes (see Figure 2–1). The Andes ranges on the eastern side are higher and geologically more complex than the older and more rounded coastal ranges to the west

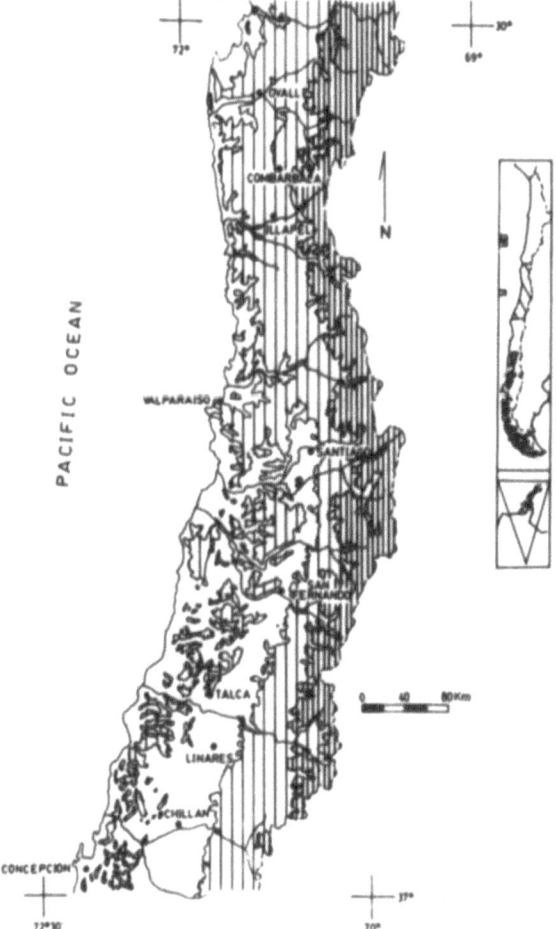

Figure 2–1. Map of the region with Mediterranean-type climate in central Chile. Notice the mountainous (dashed) vs. nonmountainous areas. Fires are associated with low altitude and are near settlements.

(Weischet 1970). Whereas the Andes can reach altitudes of 6000 m and more, the coastal ranges are rarely above 1500 m. Between these two ranges and south of Santiago there is a central valley, actually a Tertiary tectonic depression (graben), which is where most of the Chilean population lives (Fuentes 1990).

The natural vegetation in this area consists of sclerophyllous forests and woodlands (Schmithusen 1956), but since the arrival of Europeans there have been major changes in its structure and distribution (Fuentes 1990; Fuentes et al. 1990). Today, the remaining and variously modified sclerophyllous shrublands and woodlands are found on coastal ranges and on Andes ranges from about 900 m to 1500 m. On the intermediate tectonic depression, the natural vegetation has been replaced either by intensive agriculture or by an *Acacia caven* (espino) savanna (Fuentes et al. 1990).

Cities and most other human settlements are located on the intermediate depression, near large rivers capable of providing enough water even during dry years (Fuentes 1990). The wildfires that we are discussing are ignited by humans and tend to occur on the slopes of the two ranges, frequently near human settlements and on Sunday between noon and 4 in the afternoon (Avila et al. 1988), when they are associated with picnics. However, sometimes goat herders and people living at subsistence level in the foothills of the two mountain ranges ignite fires to improve the rangelands or eliminate waste products from their small-scale agricultural activities.

These are the climatic, topographic, vegetative, and cultural conditions under which fires occur today in central Chile. A first and obvious question is whether this is also a good description of the recent past. There is some evidence that the Mapuches, as the aborigines called themselves, used fires for agricultural purposes (Aschmann and Bahre 1977), but these seem to have occurred mostly near waterways. They might also have burned larger areas contiguous to these waterways to improve rangeland conditions, but that is not known for sure. But even if they burned larger areas, the total human contact with the sclerophyllous forests has not been longer than 11,000 years (Montané 1968).

Does the Chilean Matorral Really Have a Fire-Free History?

Fuentes and Espinoza (1986) provided evidence that the Chilean matorral might have been subjected to fires related not to summer lightning but to volcanic eruptions. In fact, in central Chile there are more than 20 active volcanoes, many of which are known to have erupted in ancient times. There are ashes and lavas that indicate that these Andean Quaternary volcanoes have had an effect even on the coastal ranges all the way to the other side of the central valley. Volcanic ash and lava are known to produce fires even today in the forests at the southern limit of the Mediterranean-type vegetation, and there is no reason to believe that this was different in the recent past. Pollen cores obtained by Heusser (1983) in Tagua-Tagua, in the center of the current distribution of the Chilean matorral, indicate the presence of charcoal, and thus of fires, about 40,000 years ago. The oldest records indicating the presence of humans in the region are also from Tagua-Tagua and suggest that they inhabited the area for not longer than 11,000 years. That is, there were fires for at least 30,000 years before the arrival of humans into the area, probably caused by the frequent volcanic eruptions on the Andes.

The existence of these fires, in spite of their relative paucity in comparison with the frequency of fires produced by summer lightning, could in turn explain the presence of lignotubers in the early development of many of the matorral shrubs (Montenegro et al. 1983) and some of the attributes in relation to fire that we discuss later in this chapter. However, the presence of lignotubers could also be a response to environmental stress provoked by cold or drought (Mooney 1977; James 1984) or herbivory (Fuentes et al. 1990).

At any rate, there is evidence pointing to the existence of nonhuman-related fires in the past history of the matorral, and thus toward the possible value of also using tools from evolutionary biology in interpreting some of their attributes. In the rest of this chapter we use this evolutionary scheme to account for some of the responses of matorral shrubs to fire. We initially address the resprouting and the germination responses of matorral shrubs, and then comment on the possible "strategic" meaning they might have, and on the landscape ecological implications involved.

The Resprouting Response of Shrubs

Several authors (Altieri and Rodríguez 1974; Araya and Avila 1981; Fuentes et al., personal communication) have measured the resprouting response of Chilean matorral shrubs (Figure 2–2). Essentially these results all indicate that matorral shrubs respond to fire, as to coppicing, by resprouting. However, not all species respond with the same vigor. Species such as *Lithrea caustica* (litre) and *Quillaja saponaria* (quillay), known to have deep roots (Kummerow et al. 1981), have stronger responses than more shallow-rooted species such as *Trevoa trinervis* (tevo) and *Baccharis linearis* (romerillo). Thus, it seems that species with larger underground reserves have stronger canopy flushes than species having lower root to shoot ratios (Segura et al., in press).

In addition, it has been shown (Segura et al. 1994) that the strength of this individual canopy response is positively correlated with the frequency with which individuals in the population resprout after a given fire. Species that frequently resprout after fires are also the ones that show the more vigorous resprouting responses. Again, these are attributes to be expected more frequently in species with relatively high underground reserves than in species with low underground biomass.

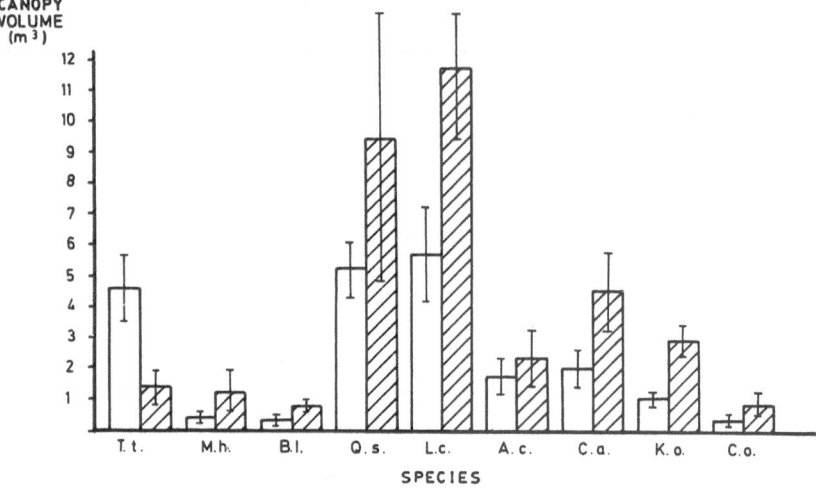

Figure 2–2. Resprouting response to fire of matorral shrubs. □ = volume of nonresprouted shoots. ▨ = volume of resprouted shoots. T.t., *Trevoa trinervis*; M.h., *Muehlenbeckia hastulata*; B.l., *Baccharis linearis*; Q.s., *Quillaja saponaria*; L.c., *Lithrea caustica*; A.c., *Acacia caven*; C.a., *Cryptocarya alba*; K.o., *Kageneckia oblonga*; C.o., *Colliguaya odorifera*. (Modified from Araya and Avila 1981.)

The Germination Response

There are reports indicating that not only in California (Christensen and Muller 1975; Keeley 1977, 1984; Keeley and Keeley 1984), but also in the other Mediterranean-type regions, there are shrub or tree species that exhibit enhanced germination after fires (Kruger 1983; Keeley 1986). This response has been interpreted as an adaptation to the recurring fires known to occur in these areas for extended periods of time.

Following the hypothesis of a long history of volcanic eruptions in central Chile, Muñoz and Fuentes (1989) tested the effect of fires to evaluate the existence of enhanced germination of at least some shrub species after fires in the Chilean matorral. They found that when the standard heat treatment in the laboratory (review by Keeley 1987) was given to a collection of seven matorral species, only *Colliguaya odifera* (colliguay), *Muehlenbeckia hastulata* (quilo), and tevo showed positive responses, whereas other species such as quillay, *Cryptocarya alba* (peumo), *Colletia spinosa* (crucero), and *Schinus polygamus* (huingan) did not have enhanced germination.

These results suggested that, as expected from the fire history hypothesis, some matorral plants could be enhanced by fire and use this disturbance for population recruitment. However, when the researchers measured the temperatures of experimental calcinating (ashing) fires in the field, they found that the conditions of the standard test (temperatures close to 100°C for 5 to 10 minutes) were not available where the soil seed bank is found. In fact, in the first 5 cm of soil where the highest proportion of the seed bank is located, temperatures are considerably higher, sometimes for hours, suggesting that the standard test might not be appropriate.

In support of this claim, the researchers found that all experimental seeds in the first 5 cm of soil were killed by the fires. This evidence indicates that calcinating fires do not have a germination enhancing capacity. It is important to realize that the fires showed to enhance seed germination in the California chaparral are precisely of this calcinating type. But not all fires in the matorral are hot enough and uniform enough to ash all shrubs in a slope. Fires frequently have a patchy effect, calcinating some shrubs and merely burning the leaves, twigs, and small branches of other shrubs. The effect of these fires seems to be different from that of calcinating ones (Segura et al. 1994).

In fact, when seeds obtained from under the canopy of completely burned shrubs (phantoms) and those from under the canopy of lightly burned shrubs were tested for viability, the seeds found under lightly burned shrubs had statistically higher viabilities than those found under the phantoms. Moreover, the seeds found under lightly burned shrubs had higher germination rates than those of the controls. These species with higher rates were tevo and quilo, which supports the previous laboratory results of Muñoz and Fuentes (1989), which indicated that these two species would be stimulated in their germination by high temperatures.

Segura et al. (1994) also found that under these lightly burned plants the natural seedlings belong mostly to these same two species, and that polar-facing and equator-facing slopes have significantly different proportions of these two types of seedlings in them. In other words, the researchers found that fires tend to produce significantly different seedling shadows and that these shadows are different on polar and on equator-facing slopes. Polar-facing slopes had seedlings of tevo, litre, quilo, peumo, romerillo, and *Podanthus mitiqui* (palo negro), whereas equator-facing slopes had only seedlings of quilo and tevo. In addition, the proportion of quilo-to-tevo seedlings was about the same on equator-facing slopes and largely skewed towards tevo on polar-facing ones. The phantoms of calcinated shrubs did not have any viable seeds and did not have any seedlings growing in them, which also support previous results by Muñoz and Fuentes (1989).

The "Strategies" of Matorral Shrubs

Keeley (1986), in discussing the fire response behavior of shrubs in Mediterranean-type ecosystems, distinguished between seeders, resprouters, and facultative resprouters. Essentially, seeders require fires to germinate, resprouters can only resprout after fires, and facultative resprouters need fires to germinate and, in addition, can resprout after fires. In his discussion, Keeley (1986) indicates that these are three strategies that would be favored by different fire frequencies and drought conditions. He ultimately sees an evolutionary and "adaptive" explanation for these three phenotypes.

How do these categories apply to central Chile? Most Chilean shrubs have the capacity to resprout after fires. Some of them are strong and vigorous resprouters. Litre, quillay, and *Schinus latifolius* (molle) are in this category. Others respond less frequently and less vigorously – tevo, romerillo, quilo, *Peumus boldus* (boldo), and colliquay, for example. This is a response they exhibit not only after fires, but also after coppicing.

On the other hand, none of the Chilean shrubs tested so far shows enhanced germination after hot, calcinating fires. This is an aspect in which Chilean and other Mediterranean-type ecosystems might differ (Keeley 1986). The increased germination rate was found in two species of weak resprouters, but only under the light burning conditions occurring in patchy fires. These two shrubs, quilo and tevo, seem to fall into a different group than the others, but they do not correspond to any of the categories defined for the California chaparral. One reason for this lack of correspondence is that they do not germinate after hot fires; the other reason is that they do not require fires to germinate. All Chilean matorral shrubs can germinate in fire-free conditions (Fuentes et al. 1984, 1986; Fuentes and Guiñez, personal communication; Lazo and Fuentes, personal communication).

Therefore, it seems that under the volcanic-driven conditions of central Chile a different suite of strategies could have been selected. Or, and this is more likely at present, it could be that lightly burned fires, which remove the canopy and create hotter and more luminous conditions at ground level, are not very different – from the plant's perspective – from other disturbances producing the same effect. In this latter case, quilo and tevo would merely be colonizers favored under disturbances large enough to produce the hot ground conditions, as simulated by Muñoz and Fuentes (1989) in the laboratory and by the light burning fires reported by Segura et al. (1994). Field experience indicates that romerillo, quilo, and tevo are frequent colonizers of roadcuts (Guiñez and Fuentes, personal communication), and of open field situations in which soil temperatures can be very high (del Pozo et al. 1989).

The remaining species, not favored by these disturbances, have other regeneration niches (Grubb 1977) requiring the shade of nurse plants to facilitate their germination and seedling survival (Armesto and Pickett 1985; Fuentes et al. 1984, 1986, 1990). Thus, the two groups reflect very general differences in the responses of Chilean shrubs to disturbance and not specific adaptations to fire.

Further research should indicate which one of these two – but not mutually exclusive – explanations deserves more credibility. New findings will be relevant not only in Chile but perhaps also in other Mediterranean-type ecosystems in which fires as well as other disturbance sources are important.

Landscape-Level Responses to Fire

With or without a causal relationship between the disturbances just described, which favor the recruitment of species such as tevo and quilo, and the long history of fires associated with volcanism in central Chile, the current effect of fires on the matorral is to modify species diversity at the landscape level. Fires, by having a patchy effect on the landscape, can favor the establishment of mixed stands of tevo, quilo, and the vigorous resprouters such as litre and quillay.

By opening large spaces to be colonized by species such as tevo, quilo, and romerillo, fires favor at least the temporal coexistence of these patches and of unburned or completely calcinated ones in which the vigorous resprouters dominate.

It is tempting to suggest that the high relative cover of species such as litre and quillay in central Chile (Mooney 1977) is already the product of a fire history. Unfortunately, this hypothesis cannot be tested using past records because they do not have the requisite accuracy. New permanent plots have to be established to test how the relative abundance of these species increases with the frequency of hot fires.

Fires can increase and reduce species diversity. A reduction is to be expected if fires are too frequent and do not allow enough time for seedlings to reach sexual maturity or for resprouters to recover the lost resources during the canopy build-up. High-frequency fires are thus expected to select for species capable of tolerating them, and therefore to reduce local species diversity. In sum, fires can increase and decrease species diversity and modify the stand structure. More research, especially long-term, should either eliminate or support these hypotheses. The evidence suggests that matorral responses are not very similar to those exhibited by other Mediterranean-type ecosystems with well-established fire histories.

Acknowledgment. This research was financed by a grant from FONDECYT.

References

Altieri, M.A., Rodríguez, J.A. 1974. Acción ecológica del fuego en el matorral natural mediterráneo de Chile, en Rinconada de Maipú. Tesis, Facultad de Ciencias Agrarias y Forestales, Universidad de Chile.

Araya, S., Avila, G. 1981. Rebrote de arbustos afectados por el fuego en el matorral chileno. *Anales del Museo de Historia Natural de Valparaíso, Chile* 14:107–113.

Armesto, J., Pickett, S.T.A. 1985. A mechanistic approach to the study of succession in the Chilean matorral. *Rev. Chil. Hist. Nat.* 58:9–17.

Aschmann, H. 1973. Distribution and peculiarity of Mediterranean ecosystems, pp. 11–19. *In* F. di Castri and H.A. Mooney (eds.), *Mediterranean-Type Ecosystems. Origin and Structure.* Springer-Verlag, Berlin.

Aschmann, H., Bahre, K. 1977. Man's impact on the wild landscape. *In* H. Mooney (ed.), *Convergent Evolution in Chile and California Mediterranean Climate Ecosystems.* Dowden, Hutchinson, and Ross, Stroudsbourg, PA.

Avila, G., Montenegro, G., Aljaro, M.E. 1988. Incendios en la vegetación mediterrànea, pp. 81–88. *In* E.R. Fuentes and S. Prenafeta (eds.), *Ecología del paisaje en Chile central. Estudios sobre sus espacios montañosos.* Ediciones Universidad Católica de Chile, Santiago.

Christensen, N.L., Muller, C.H. 1975. Effects of fire on factors controlling plant growth in *Adenostoma* chaparral. *Ecological Monographs* 45:29–55.

del Pozo, A., Fuentes, E.R., Hajek, E.R., Molina, J. 1989. Microclima y manchones de vegetación. *Rev. Chil. Hist. Nat.* 62:85–94.

Fuentes, E.R. 1990. Landscape change in Mediterranean-type habitats of Chile: Patterns and processes. *In* I.S. Zonneveld and R.T.T. Forman (eds.), *Trends in Landscape Ecology.* Springer-Verlag, New York.

Fuentes, E.R., Avilés, R., Segura, A. 1990. The natural vegetation of heavily man-transformed landscape: The savanna of central Chile. *Interciencia* 15(5): 293–295.

Fuentes, E.R., Espinoza, G. 1986. Resilience of Central Chile shrublands: A vulcanism-related hypothesis. *Interciencia* 11:164–165.

Fuentes, E.R., Hoffmann, A.J., Poiani, A., Alliende, M.C. 1986. Vegetation change in large clearing: Patterns in the Chilean matorral. *Oecologia* (Berlin) 68:358–366.

Fuentes, E.R., Otaiza, R.D., Alliende, M.C., Hoffmann, A.J., Poiani, A. 1984. Shrubs clumps in the Chilean matorral vegetation: Structure and possible maintenance mechanisms. *Oecologia* (Berlin) 42:405–411.

Grubb, P.J. 1977. The maintenance of species richness in plant communities: The importance of regeneration niche. *Biol. Rev.* 52:107–145.

Heusser 1983. Quaternary pollen record from Laguna de Tagua-Tagua, Chile. *Science* 219:1429–1431.

James, S. 1984. Lignotubers and burls – their structure, function and ecological significance in Mediterranean ecosystems. *The Botanical Review* 50(3): 225–266.

Keeley, J.E. 1977. Seed production, seed population after fire for two congeneric pairs of sprouting and nonsprouting chaparral shrubs. *Ecology* 58:820–829.

Keeley, J.E. 1984. Factors affecting germination of chaparral seeds. *Bull. Southern California Acad. Sci.* 83:113–120.

Keeley, J.E. 1986. Seed germination patterns of *Salvia mellifera* in fire-prone environments. *Oecologia* (Berlin) 71:1–5.

Keeley, J.E. 1987. Role of fire in seed germination of woody taxa in California chaparral. *Ecology* 68:434–443.

Keeley, J.E., Keeley, S.C. 1984. Postfire recovery of California coastal sage scrub. *Am. Midl. Nat.* 111:105–117.

Kruger, F.J. 1983. Plant community diversity and dynamics in relation to fire, pp. 446–472. *In* F.J. Kruger, D.T. Mitchell, and J.U.M. Jarvis (eds.), *Mediter-ranean-Type Ecosystems. The Role of Nutrients.* Springer, New York.

Kummerow, J., Montenegro, G., Krause, D. 1981. Biomass, phenology, and growth. *In* P.C. Miller (ed.), *Resource Use by Chaparral and Matorral. A Comparison of Vegetation Function in Two Mediterranean-Type Ecosystems.* Springer, New York.

Montané, M.J. 1968. Paleo-Indian remains from Laguna Tagua-Tagua, central Chile. *Science* 161:1137–1138.

Montenegro, G., Avila, G., Schatte, P. 1983. Presence and development of lignotubers in shrubs of the Chilean matorral. *Canadian Journal of Botany* 61(6):1804–1808.

Mooney, H. 1977. *Convergent Evolution in Chile and California Mediterranean Climate Ecosystems*. Dowden, Hutchinson, and Ross, Stroudsbourg, PA.

Muñoz, M., Fuentes, E.R. 1989. Does fire induce shrub germination in the Chilean matorral? *Oikos* 56:177–181.

Schmithusen, J. 1956. Die Räumliche Ordung der Chilenischen Vegetation. *Bonner Geographische Abhandlangen* 17:1–86.

Segura, A.M., Holmgren, M., Anabalón, J., Fuentes, E.R. 1994. The significance of local patchiness in the burning intensity: The case of the Chilean matorral. Oikos (submitted)

Weischet, W. 1970. Chile: Seine Landerkundliche Individualität und Struktur. Wissenschaftliche Buchgesellschaft, Darmstadt, Germany.

3. Fire Intensity as a Determinant Factor of Postfire Plant Recovery in Southern California Chaparral

José M. Moreno and Walter C. Oechel

Chaparral is an evergreen, sclerophyllous, and highly flammable vegetation that occupies large areas of California, Baja California, and parts of Arizona (Hanes 1977; Keeley and Keeley 1988). Natural and human-made fires are a recurrent phenomenon in chaparral (Byrne et al. 1977; Keeley 1982). Chaparral ecosystems are composed of a variety of shrub species with different regeneration modes after fire. It is thought that plant characteristics and life-cycle traits of many species in this ecosystem have evolved, in part, as a result of the selective pressure of fire (Hanes 1977; Keeley and Keeley 1988).

Postfire regeneration types in chaparral range in a continuum of possibilities from species regenerating only from stored seeds (obligate seeder species) to species regenerating only from resprouting (obligate resprouters). Between these extremes there are species that are capable of doing both (resprouters/seeders, also commonly known as facultative seeders) (Keeley 1977; Keeley and Zedler 1978). Questions of particular relevance are: How is species diversity maintained in communities dominated by species with different regeneration types, such as obligate seeders and facultative seeders, and what is the role of fire in species coexistence (Keeley 1977; Frazer and Davis 1988; Thomas and Davis 1989)? Coexistence of species of the obligate seeding type with those capable of regenerating from seeds and from resprouting requires that the balance between postfire adult survival and seedling production and survival of the resprouter does not offset the seedling production and survival and hence seedling production, as well as adult survival and resprouting success may be more closely related to fire, and particularly to fire intensity.

However, it is likely that resprouts will outcompete seedlings. Consequently, seedling establishment is more favorable under conditions with low postfire resprouting. Based on this reasoning Keeley and Zedler (1978) hypothesized that obligate seeders would be favored under high fire intensity conditions, whereas low fire intensity may favor resprouters. Therefore, to fully understand mixed chaparral dynamics it is necessary to know how fire affects seedling and resprout recruitment.

Resprouting success may depend primarily on the mortality of adult plants caused by the fire. Seedling recruitment of both, obligate and facultative seeding species, is practically restricted to the first or second year following a fire (Horton and Kraebel 1955; Keeley and Zedler 1978; Keeley and Keeley 1981). Seedling recruitment depends on the number of seeds present in the soil at the time of fire, their survival during the fire, postfire germination rate and subsequent seedling survival. Maintenance of obligate seeders requires that, on a relative basis, these species recruit greater number of seedlings than their resprouter counterparts by any of the previously mentioned mechanisms, and that this higher number compensates for deaths caused by competition with resprouts. Out of these factors, seed accumulation and seedling survival may interact with the fire but may not be strictly dependent on it. Seed survival and hence seedling production, as well as adult survival and resprouting success may be more closely related to fire, and particularly to fire intensity.

In this chapter we summarize a research project that was primarily designed to answer the question of what the main controlling factors are on the postfire regeneration of a southern California chaparral dominated by a facultative seeder, *Adenostoma fasciculatum*, and by an obligate seeder, *Ceanothus greggii*. (From here on they will be referred to as *Adenostoma* and *Ceanothus*, respectively.) In particular, we were interested in knowing how fire intensity (i.e., total energy release) could directly or indirectly affect seedling production, adult survival, resprouting capacity and resprout growth. The interaction between fire intensity and herbivory on resprouting of *Adenostoma* was also investigated. During the course of this project, methods were developed to estimate fire intensity after a fire. This research has been carried out at the San Diego State University's Sky Oaks Biological Field Station (33°21'N, 116°34'W, 1385 m of elevation), San Diego County, CA.

Fire Intensity and Shrub Germination

One of the most thoroughly documented features of chaparral fire-research is the dependence of seeds of many shrub and herb species on fire related cues to germinate (Keeley et al. 1985; Keeley and Pizzorno 1986; Keeley 1987; Keeley and Keeley 1987). Consequently, recruitment of shrub seedlings is practically restricted to the first or second year following the fire. Seeds that germinate between fire cycles have practically no chance of producing seedlings that survive for long periods of time, because they succumb to the action of herbivores and/or to competition with adult shrubs (Christensen and Muller 1975; Davis and Mooney 1985; Swanck and Oechel 1991). The advantages of selecting a mechanism that triggers germination after fire are many because postfire environments provide better conditions for seedling establishment, such as greater light and soil resources and, possibly, reduced competition from the surviving shrubs. However, making germination dependent on fire may be risky because too intense fire may be lethal.

The proven fact that many chaparral shrub and herbs depend on fire-related cues to germinate has somehow led to overlooking the possible detrimental effect of fire on seedling production (Keeley 1977; Zammit and Zedler 1988). Since it is likely that fire may have acted very strongly on seed traits by selecting seeds capable of surviving the fire, it is also likely that chaparral species would exhibit a differential response to fire intensity. This could be achieved through differential resistance of the seeds to heat and/or through differential distribution of them in the soil. Because obligate seeding species rely solely on their seeds to regenerate after the fire, one can, therefore, hypothesize that obligate seeders, like *Ceanothus*, should be more resistant to elevated fire intensity than facultative seeders, like *Adenostoma*.

The above mentioned hypothesis has been confirmed by our work: The two dominant shrubs of our chaparral responded very differently to increasing fire intensities (Moreno and Oechel 1991a). Athough the germination of both species was stimulated by a lower-intensity control winter fire, confirming the positive role of fire in triggering germination, *Adenostoma* greatly decreased its seedling production in the field as fire intensity increased, whereas *Ceanothus* increased its germination as fire intensity was elevated. Even at very high fire intensities, which were obtained by adding approximately twice the amount of standing biomass and litter to the normal fuel load, *Ceanothus* had germination densities similar to the control fire. The differential response of both species to increasing fire intensity can be clearly seen when expressing the germination results on relative basis to the overall shrub seedling population (Figure 3–1). At zero intensities (clip treatment) and low (control burn) fire intensities, *Adenostoma* overwhelmly dominated the postdisturbance seedling population. As fire intensity increased, *Ceanothus* became more dominant and was the dominant seedling species in the high (Bu4) and very high (Bu8) intensity treatments.

Fire intensity patterns may be operating within a burn at a spatial scale determined by fuel distribution and/or other factors (Davis et al. 1989).

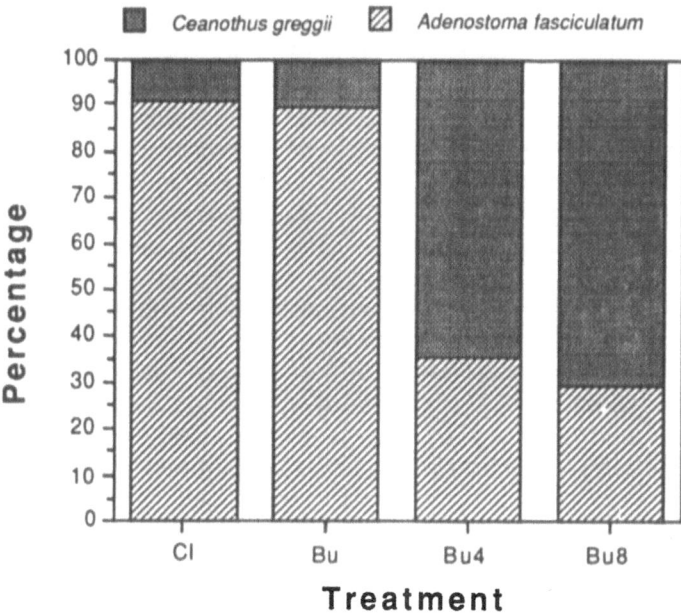

Figure 3–1. Percentage of seedlings of either *Adenostoma fasciculatum* or *Ceanothus greggii* out of the total seedling population of both shrubs seedling germinating in the field after various fire intensity treatments. Treatments were: Cl, clipped brush; Bu, control burn; Bu4, burn with addition of $4\,kg\,m^{-2}$ of brush (high intensity); Bu8, burn with addition of $8\,kg\,m^{-2}$ of brush (very high intensity. (Modified from Moreno and Oechel 1991a.)

To determine whether plot scale variations in fire intensity played a role in our system, we correlated postfire field germination with the mean minimum diameter of the stems and branches of *Adenostoma* that remain after the fire, which has been demonstrated to be a good measure of the intensity of the fire (Moreno and Oechel 1989). As expected, within the same stand and fire type we found a statistically significant negative correlation between postfire seedling density and the measure of fire intensity for *Adenostoma*, whereas no significant relationship was found for *Ceanothus* (Moreno and Oechel 1991a). Postfire germination of herbaceous species was also significantly negatively correlated with fire intensity. Furthermore, germination of *Adenostoma* was negatively correlated with prefire total shrub cover, with prefire cover of *Ceanothus*, and even with its own prefire cover (Moreno and Oechel 1992). This probably reflects its high sensitivity to fire intensity, although other factors such as litter quantity and quality may also play a role. The finding that postfire germination of *Adenostoma* may be negatively correlated with its own prefire cover is striking because *Adenostoma* accumulates its seeds beneath its own canopy (Zammit and Zedler 1988), since it has no active seed dispersal mechanism. Therefore, *Adenostoma* accumulates its genetic regeneration mechanism as well as its vegetative regeneration mechanism (resprouting, see later for the effects of fire intensity on resprouting) in places that may be sensitive to its own fuel and where it may be more susceptible to fire due to its sensitivity to increased fire intensity. Hence, *Adenostoma* has no spatial risk spreading for the two mechanisms of regeneration. On the other hand, postfire germination of *Ceanothus* was not related to any of the above prefire cover measures. *Ceanothus*, unlike *Adenostoma*, has an active seed dispersal mechanism (Evans et al. 1987) and its seeds are more widely distributed throughout the stand (Zammit and Zedler 1988).

Since resprouting species may be less adapted than obligate seeding species to the advent of Mediterranean-type climate in California (Raven and Axelrod 1978), one may argue that there might be some advantage for concentrating seeds in places where the plant had been successful. If that were the case, then seedling survival should be greater in areas where plant performance before the fire was greatest. However, when correlating first-year postfire seedling survival with measures of prefire plant abundance (either cover or density) of the same species, we found no significant relationship between these two variables for any of the two species, respectively (Moreno and Oechel 1992). Therefore, there was no apparent advantage for *Adenostoma* for concentrating both regeneration mechanisms in the same places, thus being susceptible to high fire intensity (see below for resprouting effects). Consequently, from this perspective, no compensation between seedling and resprouting mechanism is envisioned.

Postfire Time of Seedling Emergence

Postfire time of seedling emergence has not deserved much attention in chaparral research. Time of emergence may be critical, as shown for Mediterranean-type *Cistus* scrublands (Troumbis and Trabaud 1986). Zammit and Westoby (1987) studied two serotinous *Banksia* species and found that the obligate seeding *B. ericifolia* had a more gradual seed release and germination than the resprouter *B. oblongifolia*. They argued that this was an adaptation to spread the risk of germination over more than one opportunity for establishment, thus reducing the risk of zero survival. However, spreading germination may also have a negative effect if herbivory is at stake, as it will be shown later for resprouts, since chaparral seedlings are also differentially herbivorized (Mills 1983). Indeed, others have argued that synchronous seedling production in postfire environments may be advantageous to counteract herbivory, in accordance with the satiation hypothesis (Quinn 1986).

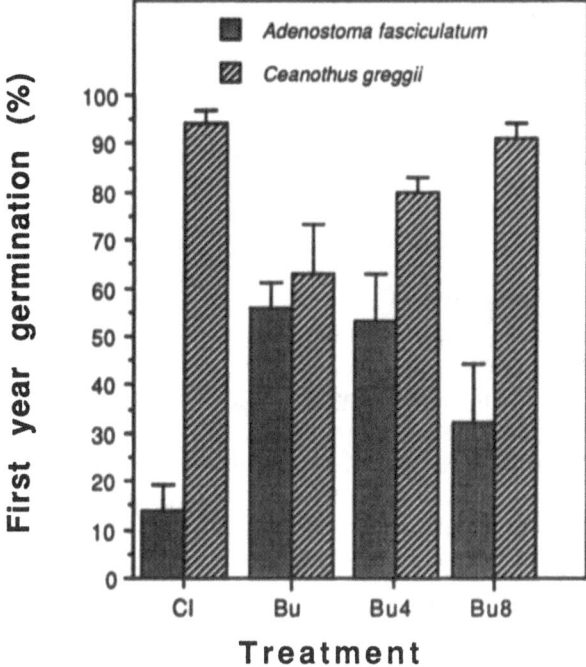

Figure 3–2. Percentage of seedlings emerging during the first year out of the total recorded during the 2 years in which germination was monitored (mean ± SE). Treatments as in Figure 3–1. (From Moreno and Oechel 1991a.)

Fire intensity could affect the time of seedling emergence by altering the soil germination environment through changes in soil temperature, litter quantity and quality, water repellency, and so forth, as well as by differentially altering the dormancy conditions of the seeds and the depth of the surviving ones. Indeed, we found that the temporal pattern of seedling emergence may be modified by the intensity of the fire (Moreno and Oechel 1991a) (Figure 3–2). At zero intensity (clip treatment) or at very high fire intensity, the proportion of germination that took place during the first year, out of the total of 2 years during which germination was monitored, was greater for *Ceanothus* than for *Adenostoma*. This latter species had a greater proportion of seedlings appearing during the second year. At the control fire intensity, both species had the most spread germination between both years. The mechanisms for these changes are not known but certainly deserve further attention. If this finding is further confirmed in other experiments and with other systems, it would point to an extension of the intermediate disturbance hypothesis (Pickett and White 1985) to another feature, namely that of seedling population diversity. Specifically, at intermediate intensities the population of seedlings may be more diverse than at lower or higher intensities.

Indirect Effects: Plant–Plant Interactions

The herbaceous flora is an important component of the postfire chaparral vegetation (Keeley at al. 1981). Herbs may compete with the shrub seedlings (Schultz et al. 1955; Kummerow et al. 1985). Hence, it is important to know their role on seedling survival because their abundance during the first year is very markedly dependent on the intensity of the fire: Increasing fire intensity reduced very significantly the number of herb seedlings, in a similar relationship to that found for *Adenostoma* (Moreno and Oechel 1991a).

To test the role of herbs on seedling establishment, seedling survival was compared between quadrats in which herbs had been removed with other quadrats in which herbs were present. This was performed for three different herb-density levels, comprising sites covering the minimum and maximum range of herb densities found in this chaparral. We found that herbs may play a role in the survival of seedlings of *Adenostoma*, but in a different way than suspected. Removing herbs tended to decrease the survival of *Adenostoma* seedlings whereas it hardly changed that of *Ceanothus* at all. This resulted in a significant difference in the survivorship

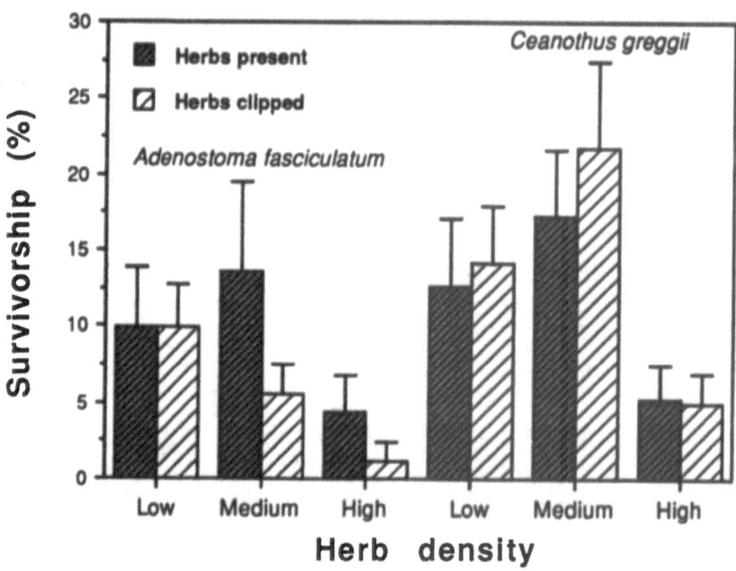

Figure 3–3. Survival of *Adenostoma fasciculatum* and *Ceanothus greggii* after the first postfire dry season in plots with herbs present or herbs removed from sites that originally had different herb densities (mean ± SE). Initial herb densities were: Low: <200 herbs m^{-2}; Medium: 201 to 400 herbs m^{-2}; High: >400 herbs m^{-2}. (From Moreno and Oechel 1992.)

of both species when herbs were removed, with *Ceanothus* surviving in greater proportion than *Adenostoma* (Moreno and Oechel 1988, 1992) (Figure 3–3).

Additionally, we also found in two different experimental fires that the germination of *Adenostoma* was positively correlated with the density of herbs, whereas *Ceanothus* germination was not correlated with herb abundance. Therefore, it appears that herbs, by altering the relative survival rates the seedlings of both species, may be beneficial to *Adenostoma*. Interestingly, this shrub tends to germinate in greater abundance where herbs do. As mentioned earlier, fire intensity reduced herb germination very markedly (Moreno and Oechel 1991a). Thus fire intensity, through its effect on herb density, may, in part, modify *Adenostoma* seedling survival. Low-intensity fires may produce greater relative germination of *Adenostoma*, and also greater number of herbs which, in turn, may enhance *Adenostoma* survival. At high fire intensity *Adenostoma* may germinate less, as will the herbs, thus contributing to a reduced survival of this shrub. *Ceanothus* will be affected by the intensity of the fire but not by the herbs in the density levels found here.

Fire Intensity and Resprouting Density

Resprouting is the main plant recovery mechanism after fire in most Mediterranean-type ecosystems (Trabaud 1987), and occurs in most mixed chaparral stands. Nevertheless, few studies have focused on the postfire demography of resprouting plants. One important problem in studying the demography of resprouting plants in wildfires is the difficulty of differentiating between prefire mortality and fire-induced mortality. Another aspect that deserves attention is the difference between direct effects on plant mortality and indirect effects. We understand by direct effects the mortality induced by the fire that causes the plants not to resprout, at least visibly at the surface of the soil. Some plants may, however, resprout but die later on as a result of other causes, which we refer to as indirect effects. To distinguish between direct and indirect effects, and to avoid any confounding effect, experimental fires have to be made in which all living plants prior to a fire must be identified, for instance, by means of a fire-resistant tag. After the fire, all tagged plants must be monitored and the status of each plant followed.

Rundel et al. (1987) have studied the postfire demography of *Adenostoma fasciculatum* in a pure *Adenostoma* stand in Sequoia National Park, CA. They provided within-fire population changes without reference to fire intensity, showing that within a single fire, mortality of mature plants decreased with increasing plant size. They also provided global

34   J.M. Moreno and W.C. Oechel

root crown mortality data for *Adenostoma* based on a qualitative estimate
of fire intensity, showing that as fire intensity increased overall plant
mortality also increased. They, however, did not distinguish from the
different causes of mortality.

At our study site, and regardless of fire intensity, direct plant mortality
was also related to plant size: the larger the plant the greater its survival
(Figure 3–4). In our case, plant size was estimated by the size of
the lignotuber, which is very highly correlated with total above-ground
biomass (Moreno and Oechel, 1993). Additionally, direct plant mortality
increased with fire intensity. In our experiment, we found 100% survival
in the clipping treatment, 96% in the control (low) fire treatment, 87% in
the high (Bu4) and 69% in the very high (Bu8) intensity treatment,
respectively (Moreno and Oechel, 1991b). In summary, at low fire intensity, plant mortality is low and restricted to low-size plants. At higher
fire intensities, greater number of plants and of increasingly larger size
succumb to the direct action of fire.

Fire intensity also affected the number of visible resprouts per unit
surface of lignotuber. Namely, the greater the intensity, the lower the
number of resprouts produced per unit surface of lignotuber (Moreno and
Oechel 1991b). Indeed, fire intensity and resprout production were
strongly negatively correlated (Figure 3–5).

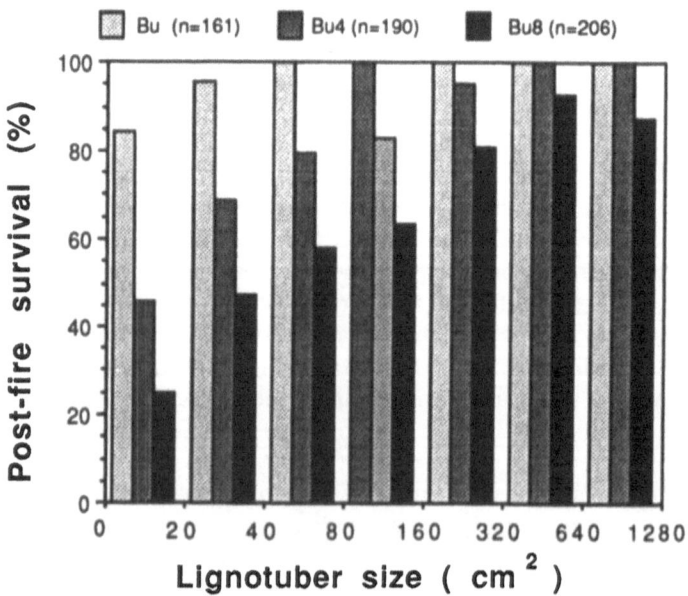

Figure 3–4. Percent survival of *Adenostoma fasciculatum* plants by lignotuber
size after fires of various intensities. Survival at zero fire intensity (clipped plants)
was 100%. Treatments as in Figure 3–1. (From Moreno and Oechel 1993.)

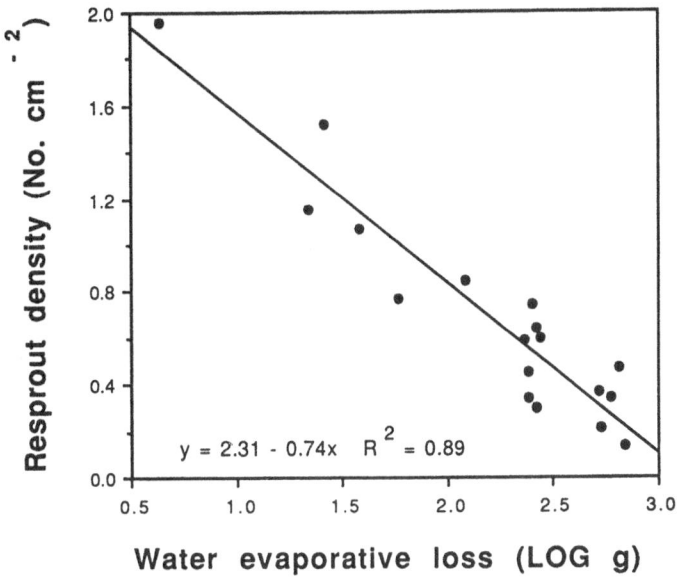

Figure 3–5. Least square regression of resprout production per unit area of lignotuber on water evaporative loss during the fire (a measure of fire intensity). Each point is the mean value of three plants. (From Moreno and Oechel, unpublished data.)

Time of Resprouting

As is the case for the seedlings, very little is known about the effect of fire on the time of resprouting. Resprouting may take place shortly after the fire or may be delayed for months, depending on the season of fire (Rundel et al. 1987; Radosevich and Conard 1980). At the plant level, Anfuso (1982) showed that south-facing parts of the lignotuber re-sprouted earlier than north-facing parts. Similarly, little is known about the time of resprouting in relation to plant size. Fire intensity may affect time of resprouting through its effects on the soil environment and/or through its interaction with the depth of the surviving buds (Noble 1985). The higher the intensity of the fire, the deeper the surviving buds may be; hence, the longer it takes the resprouts to reach the surface and the later they start to photosynthesize.

Our studies show that time of resprouting is affected by the intensity of the fire: The higher the intensity, the later the plants resprouted (Moreno and Oechel 1991b). Time of resprouting was also related, in part, to plant size, with larger-size plants resprouting earlier than small-size plants (Moreno and Oechel 1993) (Figure 3–6). This finding is harder to explain if we assume that within any one treatment, larger plants, if any, would

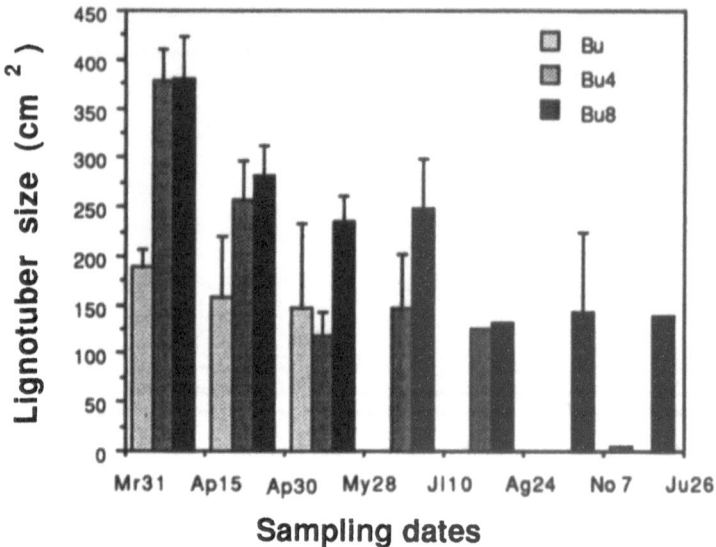

Figure 3–6. Lignotuber size of plants resprouting at various periods after a late-February fire with modified fire intensities (mean ± SE). Treatments as in Figure 3–1. (From Moreno and Oechel, unpublished data.)

burn at greater fire intensity than smaller plants due to their larger biomass accumulated right on top of the lignotuber. This fact may be explained by a larger inside growing impulse for larger plants than for smaller plants. Rundel et al. (1987) found that the slope of biomass gained during the first year after resprouting was greater for larger plants than for smaller plants. They interpreted this finding by assuming a greater relative availability of reserves by large plants than by small plants.

Indirect Effects on Resprouting

The postfire dynamics of resprouted plants during the first one or few years after fire is also a gap in our knowledge of chaparral. Little is known about how fire intensity could alter plant survival or performance. On the one hand, it is likely that soil moisture will be higher during the first dry season due to a reduction in the transpiring surface caused by the greater number of dead plants induced by higher fire intensity. On the other hand, such conditions may be more favorable for plant performance, particularly during the first postfire dry season. Therefore, lesser mortality and greater growth would take place in higher-intensity plots once they had resprouted. Nevertheless, productivity on a stand basis would, at best, be similar to low-intensity plots due to the lower number of surviving plants. It is also possible that such ameliorated conditions may make resprouts in high intensity areas more palatable and hence more susceptible to higher herbivory.

Our studies demonstrate that postfire plant mortality among resprouted plants also increased with fire intensity (Figure 3–7). This finding was related either to herbivory (see later) or to other unknown causes. Unlike predicted, mortality was also large in nonherbivorized very high intensity plants (Moreno and Oechel 1993). It is unlikely that water or nutrients were responsible for this extra number of deaths, particularly since some of the plants died during the wet season. An alternative explanation may be that the number of resprouts produced by very high-intensity plants was not large enough to provide sufficient photosynthesizing surface to produce sufficient carbohydrates to meet underground respiratory demands once stored reserves had been used. In fact, we found that within the very high fire intensity treatment, plants that died had a lower initial number of resprouts per plant than surviving plants (Figure 3–8). This may indicate that there might be a given number of resprouts per plant necessary to survive. If this number is too small, photosynthetic surface will not be restored quickly enough to be able to satisfy underground respiratory demands, thus causing the plant to die. The lignotuber

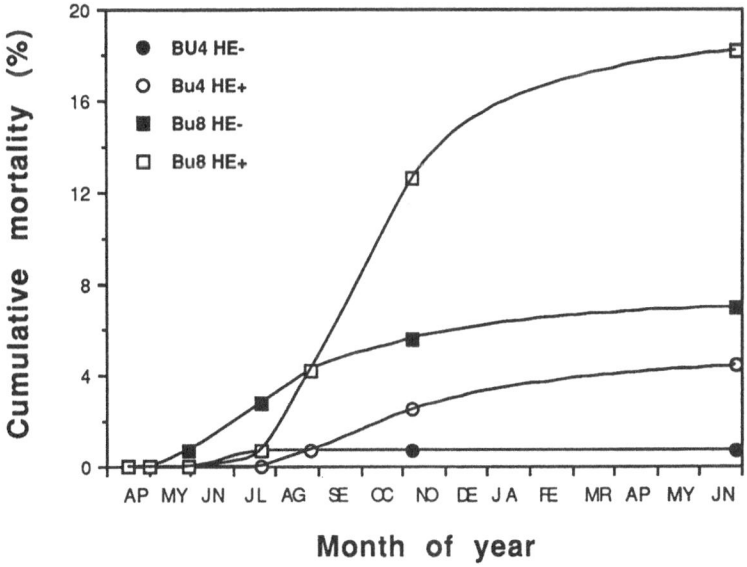

Figure 3–7. Cumulative mortality through time of plants that resprouted after an experimental fire with modified fire intensities. Mortality in the control and clip treatments was nil. Treatments as in Figure 3–1. HE[+], with herbivory, HE[−], without herbivory. (From Moreno and Oechel 1993.)

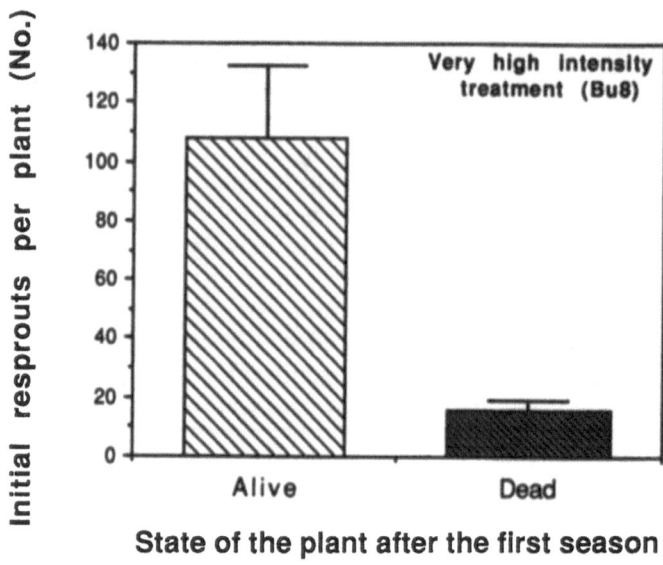

Figure 3–8. Number of initial resprouts (June 4) of plants in the very high-intensity (Bu8) treatment plots that had died or were alive about 1 year after a February experimental fire (mean ± SE). Treatments as in Figure 3–1. (From Moreno and Oechel, unpublished data.)

may then play a key role in providing enough buds to quickly restore leaf surface after a fire, allowing the plant to maintain its carbon balance.

Fire intensity interacted with herbivory, presumably by mule, deer, and/or rabbits (Davis 1967; Howe 1981; Quinn 1994), to produce very different levels of attacked plants (Moreno and Oechel 1991b). The higher the fire intensity, the greater the proportion of plants that were herbivorized and the greater also the relative intensity of herbivore attack. This means that herbivores removed most of the resprouts of the very high-intensity plants but only a few of the lower intensity plants. Differential herbivory among the various fire intensity treatments was related in part to the time of resprouting: Plants resprouting out of synchrony with the main flush of resprouting were more readily attacked by the herbivores (Moreno and Oechel 1991b). This was so regardless of the fire intensity to which plants were subjected, although it was much clearer in the very high-intensity treatment plants. As mentioned earlier, herbivory was very severe in the very high-intensity plants. We observed that many plants were repeatedly grazed to near the ground, as described also by Howe (1981). Consequently, we found that postfire mortality of resprouted plants was statistically significantly associated with herbivory. Herbivores, by eliminating the photosynthesizing surface, may have exhausted the reserves of the plants, presumably causing the plants to die.

Our studies show that, at least during the first year, fire resprouts grew better than clip resprouts. Within fire resprouts, those subjected to very high fire intensity were smaller during the first few months but by the first winter they were of similar height to control burn resprouts (Moreno and Oechel 1991b). Therefore, there was a partial compensatory response by the very high fire intensity resprouts, but these reached at best similar height to control burn resprouts. However, because the increasing intensity produced a dramatic reduction in the number of resprouts per plant, on a relative basis to lignotuber area, overall growth per plant was very much reduced. In fact, very high fire intensity plants produced nearly half the above-ground biomass per unit surface of lignotuber relative to control fire plants. Clip plants appeared also to be less productive than control fire plants (Figure 3–9). Similar results were obtained for leaf biomass: Again, very high fire intensity plants showed a reduction to nearly half of that of the control fire plants. Interestingly, the rate of

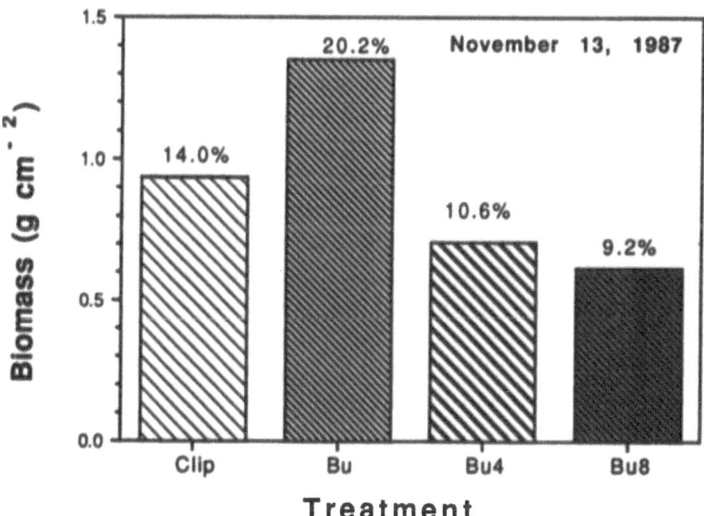

Figure 3–9. Estimated above-ground biomass per unit area of lignotuber, and percent of biomass accrued with respect to prefire levels (58-year-old plants) produced by plants by November 13, that is, 10 months after a February experimental fire with varied fire intensities. Treatments as in Figure 3–1. Productivity per unit area of lignotuber was estimated on three plants of 50 to 300 cm^2 lignotuber size per plot as follows: Total number of resprouts per plant was counted. Additionally, the height of 10 randomly selected resprouts was measured. From nearby plants, a regression was established between resprout height and biomass. From this, and based on mean resprout height and on total number of resprout per plant, biomass was calculated. Biomass of adult plants was measured from the adjacent unburned stand. (From Moreno and Oechel, unpublished data.)

recovery of foliar biomass was very rapid. One year after fire, control fire plants had recovered nearly 90% of the foliar biomass of the mature stand. Therefore, fire intensity can dramatically alter growth patterns on plants and on a stand basis, even though individual resprouts may not be so affected.

Estimating Fire Intensity After a Fire

Unlike other aspects of the fire regime that are easily determined from the date of the fire (season) or from the last fire (stand age, fire return interval), fire intensity is not as easy to estimate. This difficulty may perhaps be one of the reasons why the effects of fire intensity on eco-system properties is known in less detail. In control fires, several methods are available to estimate fire intensity. Continuous temperature recording is desirable but may be too expensive, particularly on a large scale. Temperature pellets or paints have been extensively used. These devices may still be expensive to use in large quantities.

The amount of water evaporating from cans during the fire may provide an integral measure of the fire intensity, namely an estimate of total heat release (Beaufait 1966). This measure of fire intensity may be more appropriate for studying the effects of fire on soil and plants. Water cans have been used in forest fire research to estimate fire intensity, providing excellent results (Norum 1974; Shearer 1975). To test their applicability in chaparral, we placed sets of cans containing known amounts of water adjacent to sets of temperature pellets placed at the soil surface covering a range from 100 to 900°C. Temperature pellets yield an estimate of the maximum temperature reached at the soil surface. The correlation of maximum temperatures with the water evaporated from the cans was high ($r = 0.83$) on the scale of $1 \times 1\,m$ plots in which we had one can and one set of pellets (Moreno and Oechel 1989). On a larger scale of $5 \times 5\,m$ plots, on which we had 3 sets of each, the combined results yielded a much higher correlation coefficient ($r = 0.98$). Therefore, cans filled with known amounts of water may be an excellent tool for determining fire intensity in control fires. In our case we filled the cans with 2 L of water, but this quantity may have to be adjusted to each system. Since the cans are inexpensive, this method can be used in large quantities, thus being suitable for studies on a large spatial scale to determine fire intensity patterns.

In wildfires, for which no measurements of fire temperatures or of any other type may be available, fire intensity can be qualitatively estimated after the fire. Ash type and/or depth, degree of litter or plant con-sumption, and so forth have been used, among other measures. In conifer woodland, fire intensity has been calculated from the height of scorch (DeRonde 1990). Qualitative estimates may be imprecise and subject to variation from one person to another. Quantitative, continuous, and objective estimates of fire intensity are more desirable. To determine these, we correlated the water loss from cans with measurements of the mean minimum diameter of the stems and branches of *Adenostoma* that remain unburned after the fire. The assumption was that the greater the intensity of the fire the greater the degree of consumption of stems and branches, and thus the larger the minimum diameter of the remaining branches and stems. Indeed, the correlation of branch diameters on water loss from the cans yielded a high correlation coefficient ($r = 0.83$) at the scale of $1 \times 1\,m$ plots (Moreno and Oechel 1989) (Figure 3–10): That is, the correlation of water evaporated from the cans with the mean minimum diameter of branch remains taken on all plants rooted within 1 × 1 m plots. Combining three similar measurements taken on a $5 \times 5\,m$ plot, the correlation was improved ($r = 0.95$). Therefore, measuring branch diameters of *Adenostoma* after a fire can be used to quantitatively estimate the intensity of the fire. Branch diameter measurements can be taken after a fire has occurred, presumably even months after the fire took place. This allows use of many chaparral and other shrubland fires in fire intensity studies. This approach may be operational even at scales smaller than $1 \times 1\,m$, the lowest one depending on the density of the plant that is used to measure branches and stems.

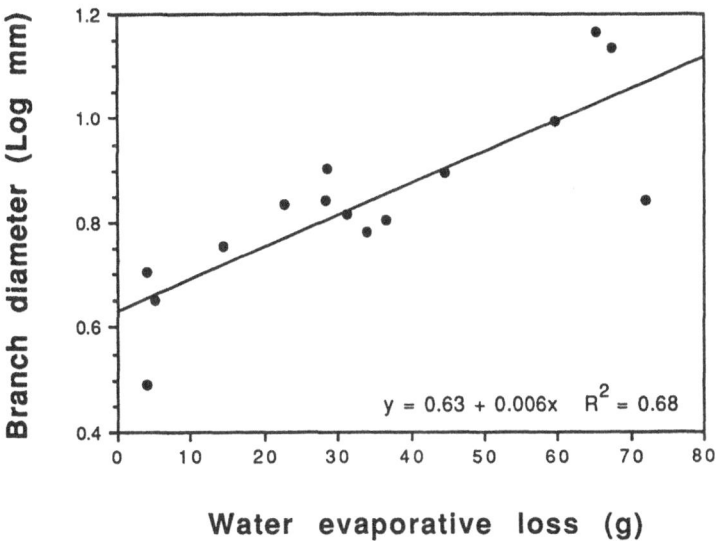

Figure 3–10. Correlation between the mean minimum diameter of branches of *Adenostoma fasciculatum* remaining after the fire and water evaporative loss during the fire. (From Moreno and Oechel 1989.)

Conclusion

Fire intensity is a crucial factor in controlling postfire dynamics of chaparral, as clearly shown by our work. Fire intensity changes with stand age, season, fuel type and density, which in turn may relate to soil depth and quality, fire weather, and topography. Even in homogeneous stands, fire intensity may vary depending on plant distribution and on the vagaries of the fire weather, producing a distinct fire pattern, as shown for grasslands (Hopkins 1965) or chaparral (Davis et al. 1989). Fire pattern is clearly visible at the landscape scale, particularly in the complex terrains of most of southern California chaparral shrublands. The complexity of effects induced by changes in fire intensity requires that this factor be fully investigated at the various space and time scales. The interaction with herbivory and other environmental factors needs to be addressed to have a comprehensive picture of chaparral dynamics and functioning.

From the perspective of species coexistence and the factors that control it, fire intensity directly or indirectly also plays a crucial role. *Adenostoma* is most favored under low fire intensity conditions. Under such conditions this species produces more seedlings, has a lower number of mature plants perishing during the burn, has survivors that produce higher number of resprouts, has plants that grow larger during the first year, and may compete better with seedlings of its own species as well as those of others. On the other hand, high-intensity conditions produce situations more favorable for the obligate seedling *Ceanothus* – greater relative germination and greater number of resprouting *Adenostoma*. Eventually, other resprouting species are killed by the fire. The survivors of resprouting species have a lower number of resprouts and grow less, which may favor seedling survival. Additionally, herb density is less which, in turn, may indirectly further favor *Ceanothus*.

Extensive fire intensity measurements should be taken in control fires to better understand fire patterns and their causes. These measurements may be expensive, but with the aid of the methods described here they should be possible without any great monetary investment. Similarly, fire intensity methods are also desirable for wildfires. Methods such as the ones described here could easily be used to obtain at least a quantitative index of the intensity of a past fire.

Our results point to important management implications. Control fires during the cool season are undoubtedly conducive to a greater dominance by *Adenostoma*. If landscape diversity is to be preserved, a program of fires combining a variety of fire intensities should be implemented.

Acknowledgment. The research here described was conducted while on leave of absence from Universidad Complutense to San Diego State University by J.M. Moreno. The work was supported by a grant of the National Science Foundation (BSR 8507699) to W.C. Oechel and NATO and Fundación del Amo fellowships to J.M. Moreno. Writing has been made possible by a grant from Universidad Complutense (Grupos Multidisciplinarios) to J.M. Moreno.

References

Anfuso, R.F. 1982. Fire temperature relationships of *A. fasciculatum*. Thesis, California State Polytechnic University, Pomona, CA.

Beaufait, W.R. 1966. An integrating device for evaluating prescribed fires. *For. Sci.* 12:27–29.

Byrne, R., Michaelsen, J., Soutar, A. 1977. Fossil charcoal as a measure of wildfire frequency in southern California: A preliminary analysis, pp. 361–367. *In* H.A. Mooney and C.E. Conrad (technical coordinators) *Proceedings of the Symposium on the Environmental Consequences of Fire and Fuel Management in Mediterranean Ecosystems*. USDA For. Serv. Gen. Tech. Rep. WO-3.

Christensen, N.L., Muller, C.H. 1975 Relative importance of factors controlling germination and seedling survival in *Adenostoma* chaparral. *Am. Midl. Nat.* 93:71–78.

Davis, F.W., Borchert, M.I., Odion, D.C. 1989. Establishment of microscale vegetation pattern in maritime chaparral after fire. *Vegetatio* 84:53–67.

Davis, J. 1967. Some effects of deer browsing on chamise sprouts after fire. *Am. Midl. Nat.* 77:234–238.

Davis, S.D., Mooney, H.A. 1985. Comparative water relations of adjacent California shrub and grassland communities. *Oecologia* 66:522–529.

DeRonde, C. 1990. Impact of prescribed fire on soil properties-comparison with wildfire effects, pp. 127–136. *In* J.G. Goldammer and M.J. Jenkins (eds.), *Fire in Ecosystem Dynamics – Mediterranean and Northern Perspectives*. SPB Academic Publishing, The Hague, The Netherlands.

Evans, R.A., Biswell, H.H., Palmquist, D.E. 1987. Seed dispersal in *Ceanothus cuneatus* and *C. leucodermis* in a Sierran oak-woodland Savanna. *Madroño* 34:283–293.

Frazer, J.M., Davis, S.D. 1988. Differential survival of chaparral seedlings during the first summer drought after wildfire. *Oecologia* 76:215–221.

Hanes, T.L. 1977. Chaparral, pp. 417–420. *In* M.G. Barbour and J. Major (eds.), *Terrestrial Vegetation of California*. Wiley, New York.

Hopkins, B. 1965. Observations on savanna burning in the Olokemegi Forest reserve, Nigeria, *J. Appl. Ecol.* 2:367–381.

Horton, J.S., Kraebel, C.J. 1955. Development of vegetation after fire in the chamise chaparral of southern California. *Ecology* 36:244–262.

Howe, G.F. 1981. Death of chamise (*Adenostoma fasciculatum*) shrubs after fire or cutting as a result of herbivore browsing. *Bull. South Calif. Acad. Sci.* 80:138–143.

Keeley, J.E. 1977. Seed production, seed populations in the soil, and seedling production after fire for two congeneric pairs of sprouting and nonsprouting chaparral shrubs. *Ecology* 58:820–829.

Keeley, J.E. 1982. Distribution of lightning- and man-caused wildfires in California, pp. 431–437. *In* C.E. Conrad and W.C. Oechel (technical coordinators). *Proceedings of the Symposium on Dynamics and Management of Mediterranean-Type Ecosystems*. USDA For. Serv. Gen. Tech. Rep. PSW-58.

Keeley, J.E. 1987. Role of fire in seed germination of woody taxa in California chaparral. *Ecology* 68:434–443.

Keeley, J.E., Keeley, S.C. 1981. Post-fire regeneration of southern California chaparral. *Am. J. Bot.* 68:524–530.

Keeley, J.E., Keeley, S.C. 1987. Role of fire in the germination of chaparral herbs and suffrutescents. *Madroño* 34:240–249.

Keeley, J.E., Keeley, S.C. 1988. Chaparral, pp. 165–207. *In* M.G. Barbour and W.D. Billings (eds.), *North American Terrestrial Vegetation*. Cambridge University Press, New York, N.Y.

Keeley, S.C., Keeley, J.E., Hutchinson, S.M., Johnson, A.W. 1981. Postfire succession of the herbaceous flora in southern California chaparral. *Ecology* 62:1608–1621.

Keeley, J.E., Morton, B.A., Pedrosa, A., Trotter, P. 1985. Role of allelopathy, heat, and charred wood in the germination of chaparral herbs and suffrutescents. *J. Ecol.* 73:445–458.

Keeley, S.C., Pizzorno, M. 1986. Charred wood stimulated germination of two fire-following herbs of the California chaparral and role of hemicellulose. *Am. J. Bot.* 73:1289–1297.

Keeley, J.E., Zedler, P.H. 1978. Reproduction of chaparral shrubs after fire: A comparison of sprouting and seeding strategies. *Am. Mid. Nat.* 99:142–161.

Kummerow, J., Ellis, B.A., Mills, J.N. 1985. Post-fire establishment of *Adenostoma fasciculatum* and *Ceanothus greggii* in southern California chaparral. *Madroño* 32:148–157.

Mills, J.N. 1983. Herbivory and seedling establishment in post-fire southern California chaparral. *Oecologia* 60:267–270.

Moreno, J.M., Oechel, W.C. 1988. Post-fire establishment of *Adenostoma fasciculatum* and *Ceanothus greggii* in a southern California chaparral: Influence of herbs and increased soil-nutrients and water, pp. 137–141. *In* F. Di Castri, C. Florest, S. Rambal and J. Roy (eds.), *Time Scales and Water Stress, Proceedings of the Vth International Conference on Mediterranean Ecosystems*. International Union of Biological Sciences, Paris, France.

Moreno, J.M., Oechel, W.C. 1989. A simple method for estimating fire intensity after a burn in California chaparral. *Oecol. Plant.* 10:57–68.

Moreno, J.M., Oechel, W.C. 1991a. Fire intensity effects on the germination of shrub and herbaceous species in southern California chaparral. *Ecology* 72:1993–2004.

Moreno, J.M., Oechel, W.C. 1991b. Fire intensity and herbivory effects on postfire resprouting of *Adenostoma fasciculatum* in southern California chaparral. *Oecologia* 85:429–433.

Moreno, J.M., Oechel, W.C. 1992. Factors controlling postfire seedling establishment in southern California chaparral. *Oecologia* 90:50–60.

Moreno, J.M., Oechel, W.C. 1993. Demography of *Adenostoma fasciculatum* after fires of different intensities in southern California chaparral. *Oecologia* 96:95–101.

Noble, J.C. 1985. Fires and emus: The population ecology of some woody plants in arid Australia, pp. 33–49. *In* J. White (ed.), *Studies on Plant Demography – A Festshrift for John L. Harper*. Academic Press Inc., London, England.

Norum, R. 1974. Smoke column height related to fire intensity. USDA For. Serv. Res. Pap. INT–157. Intermt. For. and Range., Ogden, UT.

Pickett, S.T.A., White, P.S. 1985. Patch dynamics: A synthesis, pp. 371–384. *In* S.T.A. Pickett and P.S. White (eds.), *The Ecology of Natural Disturbance and Patch Dynamics*. Academic Press Inc., Orlando.

Quinn, R. 1986. Mammalian herbivory and resilience in Mediterranean-climate ecosystems, pp. 113–128. *In* B. Dell, A.J. Hopkins and B.B. Lamont (eds.),

Resilience in Mediterranean-Type ecosystems. Dr Junk Publ. Dordrecht, The Netherlands.

Quinn, R. 1994. Animals, fire, and vertebrate herbivory in California chaparral and other Mediterranean-type ecosystems, pp. 46–78. *In* J. Moreno and W.C. Oechel (eds)., *The Role of Fire in Mediterranean-type ecosystems*. Springer-Verlag, New York.

Radosevich, S.R., Conard, S.G. 1980. Physiological control of chamise shoot growth after fire. *Am. J. Bot.* 67:1442–1447.

Raven, P.H., Axelrod, D.I. 1978. *Origin and relationships of the California flora*. University of California Publications Botany 72, Berkeley, CA.

Rundel, P.W., Baker, G.A., Parsons, D.J., Stohlgren, T.J. 1987. Postfire demography of resprouting and seedling establishment by *Adenostoma fasciculatum* in the California chaparral, pp. 575–596. *In* J.D. Tenhunen, F.M. Catarino, O.L. Lange and W.C. Oechel (eds.), *Plant Response to Stress – Functional Analysis in Mediterranean Ecosystems*. NATO Advanced Science Institute Series. Springer, New York.

Schultz, A.M., Launchbaugh, J.L., Biswell, H.H. 1955. Relationship between grass density and brush seedling survival. *Ecology* 36:226–238.

Shearer, R. 1975. Seedbed characteristics in western larch forests after prescribed burning. USDA For. Serv. Res. Pap. INT–167. pp. Intermt. For. and Range Exp. Stn., Ogden, UT.

Swanck, S.E., Oechel, W.C. 1991. Interactions among the effects of herbivory, competition, and resource limitation on chaparral herbs. *Ecology* 72:104–115.

Thomas, C.M., Davis, S.D. 1989. Recovery patterns of three chaparral shrub species after wildfire. *Oecologia* 80:309–320.

Trabaud, L. 1987. Natural and prescribed fire: survival strategies of plants and equilibrium in Mediterranean ecosystems, pp. 607–621. *In* J.D. Tenhunen, F.M. Catarino, O.L. Lange, and W.C. Oechel (eds.), *Plant Response to Stress – Functional Analysis in Mediterranean Ecosystems*. NATO Advanced Science Institute Series. Springer, New York.

Troumbis, A., Trabaud, L. 1986. Comparison of reproductive biological attributes of two *Cistus* species. *Oecol. Plant.* 7:235–250.

Zammit, C.A., Westoby, M. 1987. Seedling recruitment strategies in obligate-seeding and resprouting *Banksia* shrubs. *Ecology* 68:1984–1992.

Zammit, C., Zedler, P.H. 1988. The influence of dominant shrubs, fire, and time since fire on soil seed banks in mixed chaparral. *Vegetatio* 75:175–187.

4. Animals, Fire, and Vertebrate Herbivory in Californian Chaparral and Other Mediterranean-Type Ecosystems

Ronald D. Quinn

The plant communities of Mediterranean-type ecosystems have been the subject of many studies, but the role of animal communities in these ecosystems has received little attention. Since animals play vital roles in many ecological processes, lack of knowledge about them severely limits our understanding of community functions in these Mediterranean-type ecosystems.

Fires in Mediterranean-type ecosystems have direct and indirect effects on animal communities, just as they do in more obvious ways on plants. Fundamental properties of animal communities, such as species richness and composition, relative abundance, and energy flow can be changed for many years by a fire (Longhurst 1978; Catling and Newsome 1981; Biswell 1989; Quinn 1979, 1990). These properties are related to, and often caused by, changes in the plant community. The postfire dynamics of animal populations in Mediterranean-type ecosystems are frequently driven by shifts in the availability of various resources that stem from plant succession (Quinn 1979, 1983, 1990). Changes related to fire in the composition of animal populations can affect community processes that depend on animals, such as pollination (Moldenke 1977), maintenance of soil properties, retention and distribution of nutrients, and seed dispersal (Bullock 1978; Herrera 1984; Lloret and Zedler 1991). The first portion of this chapter reviews some direct and indirect effects of fire on animal communities in the Mediterranean-type ecosystems of the world. Where possible, comparisons are made between continents, taxa, and across time in search of common patterns (Table 4-1). This comparative approach is limited in at least three respects: (1) research on the topic as a whole is limited; (2) some Mediterranean-type ecosystems and taxa are more thoroughly studied than others; and (3) no applicable information is available from Chile, where wildfires are less frequent than in the other Mediterranean-type ecosystems (Table 4-2).

The second portion of this chapter deals with herbivory by vertebrates after fire in Californian chaparral and other Mediterranean-type ecosystems. Herbivorous animals often shape the structure and composition of regenerating plant communities through consumption of seeds, of some plant species in preference to others, and of particular plant parts (Mills 1986; Quinn 1986; Moreno and Oechel 1991b; Swanck and Oechel 1991). Herbivory can be concentrated in particular places in the plant community because of the habitat preferences of the consumers, so the impacts of herbivory are localized (Quinn 1986; Swanck and Oechel 1991).

Direct Effects of Fire on the Animal Community

Fire in Mediterranean-type ecosystems can directly destroy animals through incineration and asphyxiation. Fire kills are most common among small, relatively immobile animals that lack fire refugia. However when fires are particularly intense and fast moving, even very mobile animals such as large mammals and birds may also be killed.

Invertebrates

Populations of invertebrates living in the litter and soil were reduced by fire in the Mediterranean-type ecosystems of Australia and France by 50 to 90% (Bornemizza 1969, reported in Recher and Christensen 1981; Leonard 1974; Fox 1978; Majer 1985; Saulnier and Athias-Binche 1986; Prodon et al. 1987; Athias-Binche 1987). Similarly, in the fynbos of South Africa, grasshopper abundance was reduced by 82% immediately after fire (Schlettwein and Giliomee 1987). This drop was attributed to the destruction of two common species of wingless grasshoppers that cannot move quickly. A different pattern was observed in chaparral, in which foliage-dwelling insects were quite abundant at the beginning of the first spring after an intense wildfire (Force 1981, 1982).

Reptiles

Fire in Mediterranean-type ecosystems undoubtedly has the power to destroy reptiles by heat, although there is little published evidence that such kills frequently occur. Fire survival by reptiles is readily explained by the need of most species of reptiles to escape heat in hot, sunny Mediterranean climates. Wildfires are most likely to occur during the day and in

Table 4–1. Properties of Animal Communities in Four Mediterranean-Type Ecosystems for 5 Years after Fire

	California				Australia				Mediterranean				South Africa			
	I	H	B	M	I	H	B	M	I	H	B	M	I	H	B	M
Species persistence	+	+	+	+	+	+	+	+	+	+	+	+	+	+	+	+
Richness return	+	+	+	0	+	+	+	0	0	+	+	+	+			0
Abundance return	+	+	+	0	+	+	+	+	+	+	0	+	+			+
Postfire specialists	0	+	+		0	+	+	+			+	+				+
Patch refugia	0	+	+	+	+	+	+	+	+		+		+	+		+
Heat refugia	+	+	−	+	+	+	−	+	+	−	−		+	+		
Diet shifts							+									
References	1	2, 3	4, 5	4–8	9, 10	11, 12	13–19	8, 13 15, 18 20	21, 22	23	21	21	24	25, 26		8 26–28

Abbreviations. I, invertebrates; H, reptiles and amphibians; B, birds; M, mammals; +, yes; −, no, 0, variable; blank, no data.

Species persistence is presence before and after fire; richness return is equivalent numbers of species before and after fire; abundance return is equivalent numbers of individuals before and after fire; postfire specialists are species found only after fire; patch refugia are species that survive fire in unburned patches; heat refugia are species that use shelter to survive fires; diet shifts are species that change food patterns after fire. The first 4 properties refer to any time within the 5-year interval following fire.

References: 1, Force 1981, 1982; 2, Lillywhite 1977a, 1977b; 3, Simovitch 1979; 4, Wirtz 1982; 5, Lawrence 1966; 6, Quinn 1979, 1983, 1990; 7, Wirtz 1979, 1988; 8, Fox et al. 1986; 9, Boornemizza 1969; 10, Majer 1985; 11, Lunney and Barker 1986; 12, Christensen et al. 1981; 13, Recher and Christensen 1981; 14, Cowley 1974; 15, Fox 1978; 16, Wooler and Calver 1988; 17, Smith 1989; 18, Christensen and Kimber 1975; 19, Kimber 1974; 20, Fox 1982; 21, Prodon et al. 1987; 22, Saulinger and Athias-Binche 1986; 23, Stubbs 1985; 24, Schlettwein and Giliomee 1987; 25, Wright 1988; 26, Kreuger and Bigalke 1984; 27, Breytenbach 1987; 28, Willan and Bigalke 1982.

Table 4–2. Field Studies Cited Here, Describing Effects of Fire on Animals in Mediterranean-Type Ecosystems

Animal Class	California	Australia	South Africa	Mediterranean	Total
Invertebrates	1	8	1	1	11
Reptiles	4	4	1	1	10
Birds	5	7	0	1	13
Mammals	14	9	1	1	25

hot weather, the very times when many reptiles would be expected to be occupying heat refugia. Many of these retreats are also sheltered from the direct effects of a passing fire. I once found a Pacific Rattlesnake (*Crotalus viridis*) that had been killed by a chaparral wildfire, but this fast-moving summer fire passed the kill site just after darkness, a time when this species would normally be moving about. A few dead skinks were found after fire in the sclerophyll forests of western Australia, but many live ones were found beneath logs and stones (Christensen et al. 1981). Rock outcrops were used as fire refugia by reptiles in the sclerophyll forests of southeastern Australia (Lunney and Barker 1986). The same effect was simulated in California when caged Pacific Rattlesnakes placed in rock crevices all survived a prescribed fire in chaparral (Howard et al. 1959). Kahn (1960) concluded that the Western Fence Lizard (*Sceloporus occidentalis*) survived a chaparral wildfire by remaining beneath the soil. This species has been observed as an abundant species after chaparral fire (Lillywhite 1977a, 1977b; Wirtz 1979; Simovitch 1979). Tortoises (*Chersina angulata*) inhabiting fynbos in southern Africa survived fire by using rock crevices (Wright 1988). The young of another species of fynbos tortoise (*Psammobates geometricus*) may avoid fire by occupying habitats unlikely to burn until after the tortoises have reached reproductive maturity (Kruger and Bigalke 1984). In coastal heathlands of Greece, a population of Mediterranean Tortoises (*Testudo hermanni*) experienced population reductions at the time of fire, but quickly recovered (Stubbs et al. 1985).

Birds

Birds in Mediterranean-type ecosystems easily avoid the direct effects of all but the most intense, fast-moving fires. Flight gives them the ability to avoid approaching heat and smoke. Individual birds may return to a burned area shortly after a fire, provided that suitable habitat remains (Cowley 1974; Recher and Milledge, reported in Recher and Christensen 1981). In contrast, during fires I have seen California Thrashers (*Toxostoma redivivum*) and Rufous-sided Towhees (*Pipilo erythrophtalmus*) congregated in unusual densities in unburned chaparral shrubs at fire boundaries. These species require dense vegetative cover and would not be expected to return to the burned area.

The Nadgee fire is a dramatic example of the ability of extremely intense, rapidly moving fires to kill even the most mobile of animals. This wildfire in the heathlands and sclerophyll forests on the southeastern coast of Australia trapped and killed large numbers of birds and other animals (Fox 1978; Catling and Newsome 1981). Fox (1978) describes it as follows: "In this burst of fire, quantities of burning debris were swept up, and along with scorched and asphyxiated birds, dropped into the sea to be washed up forming a tide line, in places, 30-cm deep." There were 609 bird carcasses from 41 species picked up along that tide line, and doubtless this was only a small fraction of the total number actually killed. The number of birds within the fire area was reduced by almost half, mostly among species small in size with relatively limited flying ability. Mass bird kills by fire are exceptional in Mediterranean-type ecosystems. A team searched 0.46 km of canyon bottom after an intense chaparral wildfire and found only two bird carcasses, even though the remains of 43 mammals were discovered in the same search (Chew et al. 1959). Despite numerous searches for dead vertebrates after intense wildfires, no other instances of bird kills have been reported from Australia or California (Christensen and Kimber 1975; Wirtz 1979).

Mammals

Fire in Mediterranean-type ecosystems can immediately destroy some species of mammals, either with lethal temperatures or with toxic gases. Species that escape the killing effects of fire may be dispossessed by destruction of habitat. Some species survive fires *in situ* because they nest or rest underground, or because they can take refuge in temporary shelters below the ground as the fire approaches (Quinn 1979, 1990). Individuals that flee ahead of the flames may return to a burned home range after the fire has passed.

There is little evidence that larger mammals of Mediterranean-type ecosystems are frequently killed in fires, because they can run ahead of the flames. Unusually intense wildfires can create exceptions, just as with birds, especially when the fire is rapidly driven by high winds. The Nadgee fire killed many large mammals, trapping them between the advancing flames and the sea (Fox 1978). Most died from jumping off sea cliffs or from asphyxiation. In chaparral, a search of 0.8 ha of canyon bottom after an intense wildfire found carcasses of two large mammals (Chew et al. 1959).

Small mammals are more likely than large mammals to be killed outright by fire in Mediterranean-type ecosystems, especially if they lack access to an underground heat refuge. In chaparral, six of seven species of rodents were immediately eliminated, and presumably killed, by fire (Figure 4–1; Quinn 1979). The single species that survived, the heteromyid *Dipodomys heermanni*, was the only one of the seven species that nested

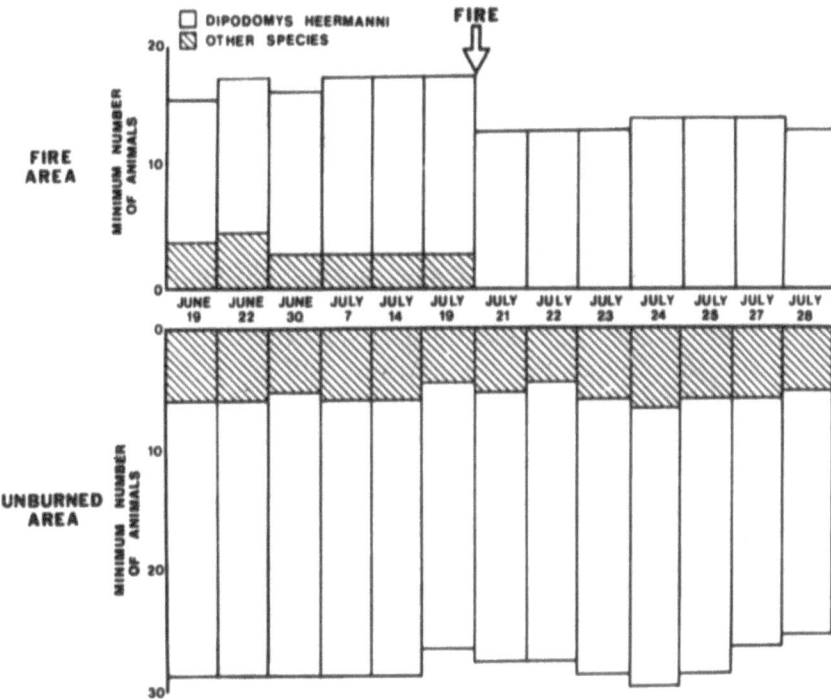

Figure 4–1. Short-term survival of an intense chaparral fire by rodents. Six species of cricetids (*Peromyscus* sp., and *Neotoma* sp.) were entirely eliminated, whereas minimum numbers of the heteromyid *Dipodomys heermanni* were not significantly reduced.

below the surface of the ground, and its population was not significantly changed by the fire. The search of a canyon bottom mentioned above found 41 dead rodents and rabbits (*Sylvilagus* spp.), all of which rest and nest above ground (Chew et al. 1959). After a search of 120 ha of hillsides following a chaparral wildfire, Wirtz (1979) concluded that populations of surface-nesting rodents (*Neotoma fuscipes*) and rabbits were destroyed by the fire. The deaths of most of the rodents were directly attributed to the fire, but most of the rabbits died from predation immediately after the fire. I have found many dead rodents and rabbits immediately following chaparral wildfires. Clearly, certain species of rodents and rabbits are frequently killed by chaparral fires. During wildfires, I have observed nocturnal species of rodents wandering about in daylight in a disoriented fashion, just ahead of advancing flames. Sometimes rabbits that have just escaped from flames turn and run directly back into a fire. Since these species of small mammals require dense vegetative cover to live, it is not surprising that they have not evolved patterns of behavior during fire that would increase their probability of immediate survival. Once their habitat has been destroyed by fire, they have virtually no chance of long-term survival on their original home ranges.

In South Africa, only two carcasses of small mammals were found after systematic searches of several fynbos fires, although a number of live animals with burn injuries were captured after these fires (Breytenbach 1987).

Postfire Successional Changes in Animal Communities

Animal communities of Mediterranean-type ecosystems respond to the fire regime in several ways. The response pattern of each type of vertebrate is similar across all Mediterranean-type ecosystems (see Table 4–1). Mammals, and quite probably reptiles, show marked changes in species composition and abundance after fire, and some of these changes may continue until the next fire (Lillywhite 1977b; Longhurst 1978; Quinn 1979, 1990). Responses of invertebrates differ according to microhabitat. Those dwelling in the litter and soil sharply decrease after fire, and the numbers of some taxa may remain below prefire levels for years (Abbott 1984). In contrast, foliage-dwelling insects, at least in chaparral, have high abundances and species richness shortly after fire (Force 1981, 1982). The avifaunas of Mediterranean-type ecosystems show remarkable resilience to fires. Abundances of some species change, but few species disappear altogether, and overall species richness commonly remains near prefire levels (Wirtz 1982).

Invertebrates

Fire in the sclerophyll forests of southwestern Australia reduced the abundance of some, but not all, taxa of soil invertebrates for at least 3 years (Abbott 1984). In these forests, prescribed fire at close intervals may permanently simplify the soil biota (Springett 1976, 1979). In contrast, Force (1981, 1982) found that foliage insects of chaparral were quite abundant in the first year after an intense wildfire. Flower-feeding species, closely followed by their insect predators and parasites, were attracted to the site by the prolific growth and flowering of vegetation that occurred after fire. Force (1981) hypothesized that the abundant insects present the first spring after fire had migrated from unburned areas 2 km distant, or from islands of unburned vegetation closer by. Fire pattern is a potentially important aspect of the fire regime for animals with limited mobility. Larger fires that thoroughly remove all vegetation and destroy animal populations might be recolonized more slowly by classes of animals with limited mobility. This includes invertebrates with weak powers of dispersal, many species of reptiles, and possibly some species of small mammals.

Reptiles

The maximum number of species and individuals of reptiles is reached in Californian chaparral in the early years after fire, when regenerating shrubs provide an optimal mixture of cover and open areas for foraging and thermoregulation (Lillywhite 1977a, 1977b; Simovich 1979). In contrast, species of lizards inhabiting *Banksia* woodlands of western Australia showed variable population responses to fire, according to the habitat requirements of each species (Bamford 1985). Some species disappeared after fire, others increased, and still others seemed unaffected; the lizard community as a whole contained about the same number of individuals at all stages of succession.

Fire can have a severe and sustained impact on some species of reptiles in Mediterranean-type ecosystems. In the coastal forests and heathlands of southeastern Australia, 2 of 18 species of lizards were eliminated by fire for at least 4 years (Lunney and Barker 1986). Tortoises inhabiting fynbos of South Africa slowly recolonized areas where fire had killed a large part of the population. This was attributed to limited dispersal ability and high rates of predation in habitat where fire had reduced vegetative cover (Wright 1988). Other species of reptiles are quite resilient to fire. After fire killed a large fraction of a tortoise population in Greek heathland, surviving immature animals grew to reproductive maturity more rapidly (Stubbs et al. 1985). In the sclerophyll woodlands of southeastern Australia, two species of scincid lizards maintained populations after fire (and drought), due to flexible foraging behavior and diet (Lunney et al. 1989).

Birds

The bird communities of most Mediterranean-type ecosystems show relatively limited, short-term changes after fire. Abundances of some species change, but species richness commonly remains near prefire values. The changes that do occur are best explained by alterations in habitat structure of the vegetation, and shifts in the availability of food resources.

Studies of the avifauna of sclerophyll forests in Australia found little change in species richness after fire (Wooler and Calver 1988; Smith 1989). Fire effects seemed to lie more in abundance changes than in the species present (Christensen and Kimber 1975; Catling and Newsome 1981; Recher and Christensen 1981; Wooler and Calver 1988; Smith 1989). In chaparral, avian species richness can increase after fire (Wirtz 1982), and there can be an increase in overall abundance of birds accompanied by a shift from species preferring dense shrubs to those that live in more open habitats (Lawrence 1966). Twelve species of resident passerines in sclerophyll forests of southwestern Australia showed a slight decrease in overall numbers the first year after an intense prescribed fire, followed by a significant increase by the second year (Kimber 1974). The increase in year 2 was accounted for among the inhabitants of the lowest two layers of vegetation, where fire had had the greatest effects. Species of birds in the higher vegetative strata were largely unaffected by the fire. The avifauna of the Mediterranean climate oak forests of southern France was relatively stable and unchanged after fire, despite short-term changes in the structure of the plant community (Prodon et al. 1987).

Diet often plays an important role in the postfire dynamics of the avian community in Mediterranean-type ecosystems. In the sclerophyll forests of southwestern Australia, abundances of the six to eight most common species of insectivorous birds living in the lowest layers of vegetation were reduced by about half for 3 years after fire (Wooler and Calver 1988). The birds that remained survived through flexibility in the array of food items consumed, shifting to a less profitable diet that supported fewer individuals. Wirtz (1982) attributed increases in populations of granivorous and insectivorous birds after a chaparral wildfire to food resources provided by the rapid and prolific regrowth of the plant community. Diet also entered into the explanation of shifts in species abundances after fire in a sclerophyll forest of southeastern Australia (Smith 1989). Birds bred outside of the normal season after that fire, which was attributed to an unseasonal increase of insects living on the flush of postfire plant growth. Similar extensions of the breeding season were observed after an intense wildfire in southeastern Australia (Recher and Christensen 1981).

A postfire study of birds in the coastal sage scrub plant community of southern California by Stanton (1986) showed a pattern quite different from that of the nearby chaparral or the other Mediterranean-type ecosystems discussed earlier. Species richness and overall abundance were higher in unburned than in burned, coastal sage. The mature vegetation met the foraging and breeding requirements of more individuals and species of birds at all times of the year than did a recently burned area.

Mammals

Of all classes of vertebrates in Mediterranean-type ecosystems, mammals undergo the greatest short-term changes in response to fire. In the months following fire, some species of small mammals completely disappear, others decrease, while still other species sharply increase. The three population cycles shown in Figure 4–2 apply to all four Mediterranean-type ecosystems where wildfires are frequent. Examples of these cycles exist for small mammals in France (Prodon et al. 1987), California (Quinn 1979, 1990; Wirtz 1982; Wirtz et al. 1988), South Africa (Breytenbach 1987), southwestern Australia (Christensen and Kimber 1975), and southeastern Australia (Fox 1982). A given community may have more than one species showing a particular type of cycle. Mammalian species represented by curve A require dense vegetation, deep litter, or some other community attribute associated with late successional vegetation. These

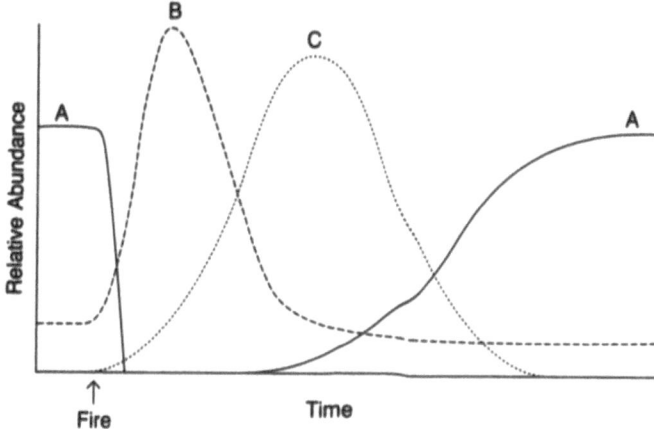

Figure 4–2. Generalized model of community dynamics of small mammals between fires in Mediterranean-type ecosystems. The time scale could vary from approximately 10^1 to 10^2 years, depending on the ecosystem and site specific fire regime. (See text for details.)

populations are destroyed or displaced during or shortly after fire, and reappear and increase in numbers only after sufficient time has elapsed for requisite development of the plant community. Population curve B shows species of small mammals that are rare at the time of fire and become much more abundant for a time following fire. Indeed, during the early stages of succession these species are often the most abundant, or only, mammal present in their body size class. Some of these are species that favor conditions caused by recent disturbance. Others are habitat generalists with high intrinsic rates of increase (r), that flourish in the absence of interspecific competition: they are the zoological equivalents of weeds. Their decline often coincides with the rise of populations represented by curve C. These are midsuccessional species and strongly associated with the rapidly regenerating plant community. Such species often are entirely absent at earlier and later successional stages. A particular community may have a family of type C curves representing species that overlap in time to varying degrees, with population maxima that do not necessarily coincide (Quinn 1979). Large mammalian herbivores of Mediterranean-type ecosystems also fit curve C. The ungulate *Odocoileus hemionus* (Mule Deer) occurs in low densities in mature chaparral, but becomes much more abundant in recently burned areas (Taber and Dasmann 1958; Biswell 1961). The regenerating plants provide easily accessible food of high quality that deer favor, especially when it is near dense vegetative cover. The same pattern applies for the macropods *Macropus fuliginosus* (Grey Kangaroo) and *M. irma* (Brush Wallaby) after fire in the sclerophyll forests of southwestern Australia (Recher and

Christensen 1981). Several African antelopes, *Pela capreolus* (Grey Rhebuck), *Oreotragus oreotragus* (Klipspringer), *Redunca fulvorufula* (Mountain Reedbuck), and the perrisiodactyl *Equus zebra* (Mountain Zebra) are similarly attracted to areas of recently burned fynbos (Kruger and Bigalke 1984). Not all populations of all species neatly fit one of these categories. Nevertheless, almost all fit one curve better than the others, and apparent differences may be attributed to variations in fire regimes. Each type represents a fundamentally different way to live successfully in an environment that changes rapidly but consistently between fires.

Almost without exception, mammalian species of fire prone Mediterranean-type ecosystems do show successional population changes. Although the amplitude of these changes is often quite large, there is no evidence that the mammalian community is destabilized or permanently shifted to a new condition by fire, unless the fire regime is altered. The changes, like those of the vegetation, are cyclical. They are driven by the fire cycle, and have periods equal to the fire interval. For a particular mammalian species, changes in population size may continue over most of a fire cycle (type A), while for other species the portion of the cycle marked by distinct changes in population size may be much shorter than the period of the fire cycle (type B). Individual species of mammals track changes in habitat quality that are linked to plant succession (Taber and Dasmann 1958; Biswell 1961; Fox 1982; Quinn 1983, 1990; Fox et al. 1986). The presence and relative abundances of species of mammals in Mediterranean-type ecosystem shrublands can be accurately predicted from specific attributes of the successional plant community such as vegetative cover, foliage height diversity, presence of shrubs, shrub height, and dense forbs and grasses near the ground (Bayless 1980; Catling et al. 1982; Price and Wasser 1984; Breytenbach 1987).

In cases in which species of mammals are entirely eliminated by a fire, they must return from populations in nearby unburned areas. The relationship between distance to a source of immigrants and the time required for reestablishment of animal populations has not been investigated for Mediterranean-type ecosystems. A source of immigrants is usually close at hand because wildfires in Mediterranean-type ecosystems often leave patches of unburned or partially burned habitat within burned areas (Fox 1978; Minnich 1989). Animal populations subject to extinction at any point in the fire cycle (types A and C) have discontinuous ranges that are a product of fire pattern. Population distributions shift in time and space as each species responds to the continuously shifting mosaic of suitable habitat created by the fire regime. In ecosystems where local extinctions occur, the rate at which a species can recolonize an area of suitable habitat can be an important aspect of its life history. Reestablishment of a mammalian population by immigration in a burned area often occurs in steps. The first immigrants may be mostly transient males, and it can take several years after the appearance of the first individuals to establish a large population of successfully breeding females (Christensen and Kimber 1975; Fox 1982; Prodon et al. 1987). This sequence of recolonization suggests that there may be a lag between the time when plant succession creates suitable habitat for a species that is absent and the time when a viable population becomes established. The duration of this time lag for a particular species is a function of its dispersal rate and of the distance to the nearest source of immigrants. The former is a property of the animal species, and the latter an attribute of recent fire history (fire regime).

Vertebrate Herbivory in Mediterranean-Type Ecosystems

Herbivory by vertebrates is an important force shaping the structure and function of Mediterranean-type ecosystems. Animals can affect the species composition of the plant community through selective consumption of seeds or seedlings. This is particularly true after fire, when most plant species must replace individuals, or even entire populations, that were killed by the fire (Mills 1983, 1986). Consumption of understory plants can modify, or indeed even eliminate, this stratum from a plant community (Christensen and Muller 1975a; Swanck and Oechel 1991). Herbivory upon foliage and branches can alter the structure of mature plants, and may even affect their ability to reproduce (Gibbens and Schultz 1962; Davis 1967). Herbivory can be concentrated at particular places in the plant community due to the habitat preferences of vertebrates, such as ecotones at the edges of mature shrublands and openings in otherwise continuous shrublands (Bartholomew 1970; Halligan 1973; Larson 1985; Swanck and Oechel 1991).

Herbivory and *Ceanothus* Succession in Californian Chaparral

Herbivory strongly affects patterns of postfire establishment of *Ceanothus crassifolius* Torr. (Rhamnaceae) in chaparral. Populations of this perennial shrub are completely killed by fire, so the next generation of plants must arise entirely from the soil seed bank. Since the genus *Ceanothus* is widespread and abundant in Californian chaparral (Hanes 1977), and reproduction after fire by obligate seeding is a characteristic of approximately half of the 50 to 60 taxa of *Ceanothus* occurring in chaparral (Wells 1969), the results described here are probably applicable to the postfire development of chaparral in many areas. Three variables can affect the number of seeds of *C. crassifolius* that germinate from the soil seed bank after fire: (1) quantity of viable seeds produced before the fire, (2) consumption or removal of seeds by animals, and (3) heat survival and scarification (Figure 4–3).

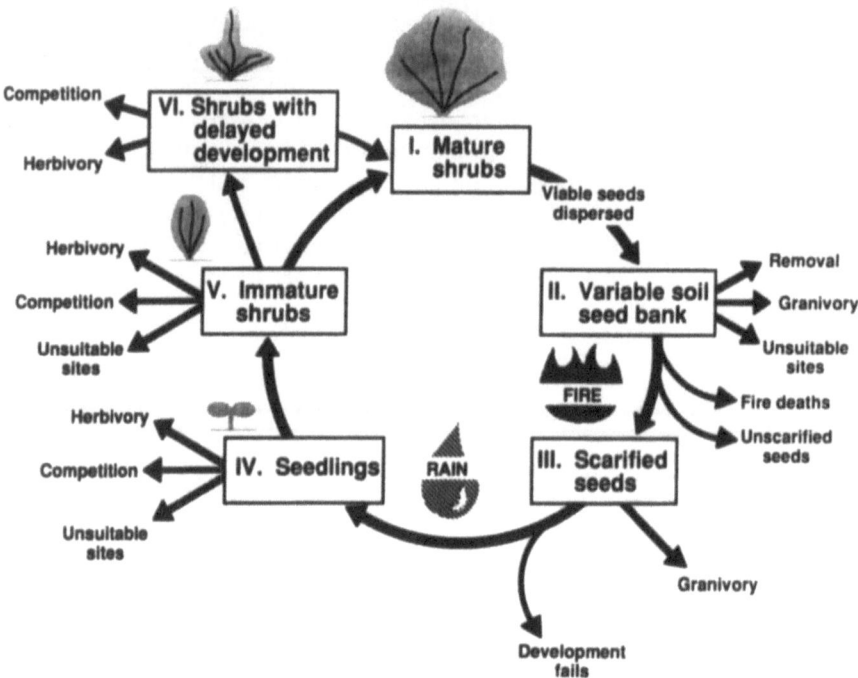

Figure 4–3. Fire cycle and life history of *Ceanothus crassifolius* in Californian chaparral. Arrows not connecting life-stage boxes (I–VI) identify sources of mortality.

Granivory and the Soil Seed Bank

The quantity of *C. crassifolius* seeds produced annually by mature shrubs varies by three orders of magnitude (Table 4–3). This degree of variation within and between populations of perennial shrubs is not unusual for Mediterranean-type ecosystems (Parker and Kelly 1989). The entire annual crop of seeds is ballistically dispersed to the ground during a period of 5 to 30 days at the beginning of summer. The density of seeds in and upon the soil immediately after dispersal is rapidly reduced by seed predators, so that within a few months most seeds of the year have been consumed or removed (Figure 4–4). Rapid depletion of soil seed banks has been observed for other species of chaparral shrubs, and in other Mediterranean-type ecosystems (Parker and Kelly 1989). Because the soil seed bank of *C. crassifolius* is subject to a highly variable annual pulse and continuous losses, seed density is never constant and there is no net accumulation of seeds with time (see Figure 4–4). Because of seed predation, the density of the soil seed bank is almost always lower than the density of the annual seed rain, a condition that seems to apply to other species of chaparral shrubs as well (Keeley and Hays 1976; Keeley

Table 4–3. Mean Density of Seeds Collected from an Array of 100 Seed Traps Placed Beneath 20 Randomly Selected *Ceanothus crassifolius* Shrubs from the Same Population

Year	Seeds m^{-2} ($\pm$SD)
1979	172 $\pm$ 112
1980	2428 $\pm$ 1146
1981	2758 $\pm$ 1078
1982	27 $\pm$ 27
1983	737 $\pm$ 775

1977, 1987; Davey 1982; Zammit and Zedler 1988; Parker and Kelly 1989).

The density of new *C. crassifolius* seeds (<1 yr of age) in the soil at a particular time depends on the size of the latest seed crop, and time elapsed since seedfall. Since *Ceanothus* seeds retain viability for many years, the soil seed bank may contain living seeds that have lain dormant in the soil for decades (Quick 1956; Quick and Quick 1961; Barro and Quinn, unpublished manuscript). The ratio of new seeds to older seeds (>1 yr of age) varies, depending on the relative sizes of the most recent seedfall and seedfalls of previous years (Figure 4–5). In 1990, when the seed crop of *C. crassifolius* was large, soil seed density immediately after seedfall was high and 92.8% of all seeds were new. In 1991, seedfall was much smaller, and only 51.5% of the subsequent soil seed bank was

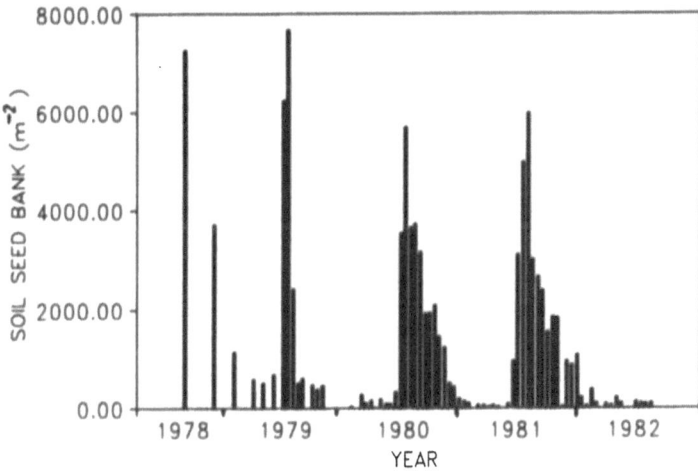

Figure 4–4. Mean density of the soil seed bank of *Ceanothus crassifolius* for 4.5 years. The peak for years 1978 to 1982 coincides with July. Each value is from 10 samples taken beneath 10 randomly selected shrubs from the same population.

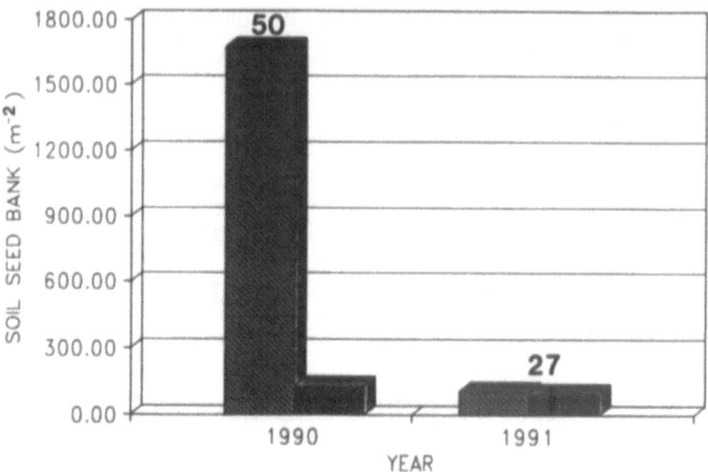

Figure 4–5. Proportion of new (diagonal shading) and old (black) seeds of *Ceanothus crassifolius* in the soil seed bank immediately after seedfall. Sample size is given above each year.

new seeds. The absolute density of old seeds was only slightly reduced between 1990 and 1991. These results suggest that carryover seeds from previous years could be important in the regeneration of *Ceanothus* when fire follows a poor seed crop. Given continuous granivory, a year of low seed production followed in the same year by fire could produce a low initial density of *Ceanothus* seedlings. Depending on the degree of seedling reduction and the pattern of subsequent shrub thinning this combination of factors could, in turn, lessen the frequency of *Ceanothus* in the ensuing community of mature shrubs.

The effectiveness of fire in initiating a new population of an obligate seeder like *C. crassifolius* depends on when the fire occurs. Wildfires can occur at any time of the year in Californian chaparral (Countryman 1974). A fire in July, just after seeds have fallen, will impinge upon the largest soil seed bank of that seed year (see Figure 4–4), and might be expected to produce a relatively high density of viable seeds that have been scarified. Later fires will affect a seed bank that becomes progressively smaller, until the next seed pulse at the beginning of the following summer. Fires in spring, 9 to 11 months after seedfall and a popular time for prescribed fires in Californian chaparral, have the potential of greatly reducing the frequency of *Ceanothus* seedlings in the regenerating plant community. The importance of the season of a fire to the subsequent abundance of *C. crassifolius* was illustrated after a wildfire in March, when no seedlings of the species appeared, even though it had been well represented on the site before that fire and before the preceding fire (Horton and Kraebel 1955). A fire in early June, almost a full year after seedfall, nearly eliminated *C. crassifolius* from a mixed species stand of chaparral (Howe and Carothers, 1980). The dynamics of the soil seed bank of this and other plant species with similarly timed life histories make them vulnerable to population reduction, or local extinction, when fire occurs in the spring. This sensitivity varies from year to year, depending on the size of the previous seed pulses (see Figures 4–4, 4–5). Conversely, summer fire would operate on larger soil seed banks, produce higher seedling densities, and potentially increase the relative frequency of *C. crassifolius* in the subsequent community of mature shrubs. Herbivory upon seedlings can modify both of these outcomes, as will be discussed in the next section.

Fire destroys many *C. crassifolius* seeds, but at the same time fire is required for germination. Heat scarifies seeds that are shallowly buried in the soil. Seeds at or near the surface are likely to be killed, while those buried too deeply will not receive sufficient heat for scarification, and will not germinate (Zammit and Zedler 1988). Germination tests in the laboratory suggest that the optimum heat range for scarification of *Ceanothus* is 85 to 90°C (Quick and Quick 1961). In high-intensity chaparral fires during summer, maximum temperatures of 90°C have been measured at soil depths of approximately 2.5 cm (DeBano et al. 1979; Quinn 1979). In a low-intensity winter chaparral fire this temperature was reached at a soil depth of 0.6 cm (Moreno and Oechel 1991a). Seeds of the Australian shrub *Acacia suaveolens* (Sm.) Willd. (Fabaceae) are similarly scarified by fire only at soil depths of 1 to 6 cm (Auld 1986). Fires of very low intensity may severely limit, or altogether eliminate, regeneration of *C. crassifolius* due to insufficient heat for scarification of buried seeds (Riggan et al. 1988).

Ten to eleven percent of all *C. crassifolius* seeds found in the soil 1 week after a wildfire in July appeared to be viable; the remainder had been killed by fire (Figure 4–6). Rodents, and perhaps birds, continued to remove edible seeds (viable and roasted) of *C. crassifolius* from the soil seed bank during the 4.3 month interval between the fire in July and the next substantial rainfall in December. During that interval, mean soil density of viable seeds was reduced from 135 to 45 m^{-2} (Figure 4–6). Virtually all viable and roasted seeds remaining in the soil during the rainy season (December to April) had either germinated or been removed by April (Figure 4–6). Within a 10 m × 15 m exclosure against all terrestrial mammals, viable seeds of *C. crassifolius* remained in densities of 45 m^{-2} at 9 months after the fire. At the same time, no viable seeds were found within a nearby exclosure that was identical, except that it admitted rodents while excluding all larger mammals. These results support the hypothesis that rodents present after fire in the burned chaparral area consumed surviving seeds of *C. crassifolius* until all seeds had either germinated or were completely removed. The viable seeds that remained within the exclosure against rodents 9 months after fire probably failed to

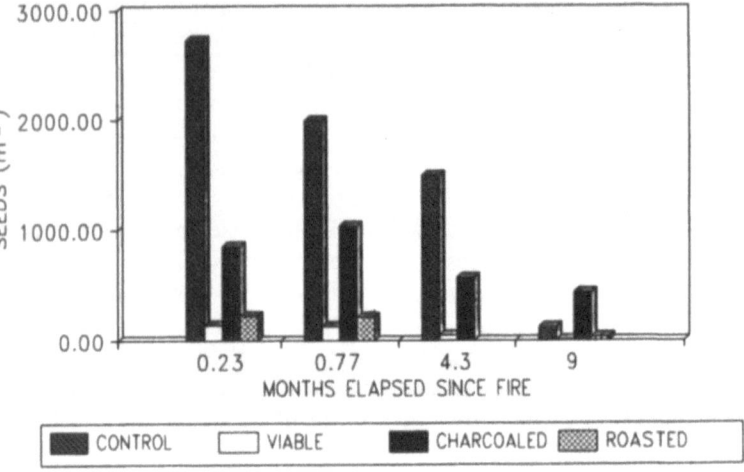

Figure 4-6. Soil seed bank of *Ceanothus crassifolius* after wildfire. All values are the mean of 10 soil samples taken randomly from beneath *C. crassifolius* shrubs. Soil samples were 22 × 17 × 10 cm deep. Control values are derived from viable seeds in an adjacent stand of unburned *C. crassifolius* chaparral.

germinate because they were buried too deeply for heat scarification. Sampling removed soil to a depth of 10 cm, well below the level where temperatures sufficient for scarification were probable (Moreno and Oechel 1991a; Quinn 1979). If this interpretation is correct, then virtually all seeds of *C. crassifolius* outside the rodent exclosure that failed to germinate, including any buried to depths up to 10 cm, were found and consumed by rodents within a period of 9 months. Live trapping of rodents during this time period revealed that the heteromyid rodent *Dipodomys agilis* was the most abundant small mammal in the burned area, occurring at minimum densities of approximately $3 \, ha^{-1}$. Animals of this genus are granivores, and are skilled at detecting and excavating seeds buried in the soil (Price and Waser 1984). Populations of *D. agilis* flourish in chaparral for the first 4 years after fire, foraging for seeds on the open ground created by the fire (Quinn 1979, 1990; Wirtz 1982). Viable seeds of *C. crassifolius* remained in the soil seed bank of adjacent mature *Ceanothus* chaparral (21 years old) during the experimental period of 9 months, while seed densities declined from $2722 \, m^{-2}$ to $119 \, m^{-2}$ (Figure 4-6). *D. agilis* and other species of rodents that eat seeds were present there also, but without destruction by fire, viable seeds were present in initial densities more than an order of magnitude larger, and it was probably more difficult for animals to find all *Ceanothus* seeds beneath mature chaparral in the physically heterogeneous environment of the soil-litter layer (Price and Reichman 1987). Moreover, *Dipodomys* forages preferentially in open areas (Price and Waser 1984).

Herbivory and Seedlings

Seedlings are especially vulnerable to herbivory; at this stage, they lack physical and chemical defenses against herbivores that may develop later, and their small size allows them to be killed by a single bite from even the smallest vertebrate. During the first growing season after fire in Mediterranean-type ecosystems, when the seedlings of many species are developing together, preferential consumption of one species over another has the power to shift the outcome of interspecific competition. When this shift occurs between the seedlings of shrub species that have the potential to become dominants, it can change the structure and species composition of the maturing plant community until the next fire. Fire not only resets the life cycle for individual plant populations (see Figure 4–3), it is a stochastic event that can alter the long-term properties of the community. Variables that can change the community between fire cycles fall into three categories: (1) variables within the fire regime, including fire intensity, season, spatial pattern, interval since the last fire, and mean interval or longest interval between all previous fires (Keeley 1977; Mooney et al. 1981; Fox and Fox 1987); (2) size of the soil seed bank, which is influenced by weather during the reproductive lifespan of plants before the fire; and (3) the presence and activities of plant predators.

Herbivory upon seedlings and young plants affects the status of *Ceanothus* in the chaparral plant community. After fire, small mammals killed a larger proportion of the *Ceanothus* seedlings in mixed populations of *C. greggii* and *Adenostoma fasciculatum* H. and A. (Rosaceae), thus favoring the establishment of *A. fasciculatum* over *C. greggii* in the regenerating plant community (Mills 1983). When seedlings of all species were protected from vertebrate herbivory with exclosures, the balance between the two species was reversed, with higher survivorship of *C. greggii* seedlings.

When vertebrate herbivores had access to burned chaparral for a decade after fire, a shrub community dominated by *Ceanothus crassifolius* before the fire became a mixed community of coastal sage scrub and chaparral, dominated by the coastal sage species *Salvia mellifera* Greene (Lamiaceae) (Figure 4–7A). This was a dramatic shift in the plant community. Coastal sage scrub generally grows on more xeric sites than chaparral, and it has a different plant species composition, physical structure, physiology, phenology, and animal inhabitants (Westman 1982; Price and Waser 1984). This shift in species composition did not occur within a nearby 10 × 15 m exclosure against vertebrate herbivores larger than rodents. *Ceanothus* there accounted for 80% of relative vegetative cover, and *S. mellifera* was almost absent (Figure 4–7B). Mammalian herbivores that could have consumed seeds, seedlings, or foliage of *Ceanothus* outside the exclosure ranged in body mass from 10 g to 50 kg, and included nine species of rodents, two species of lagomorphs

A

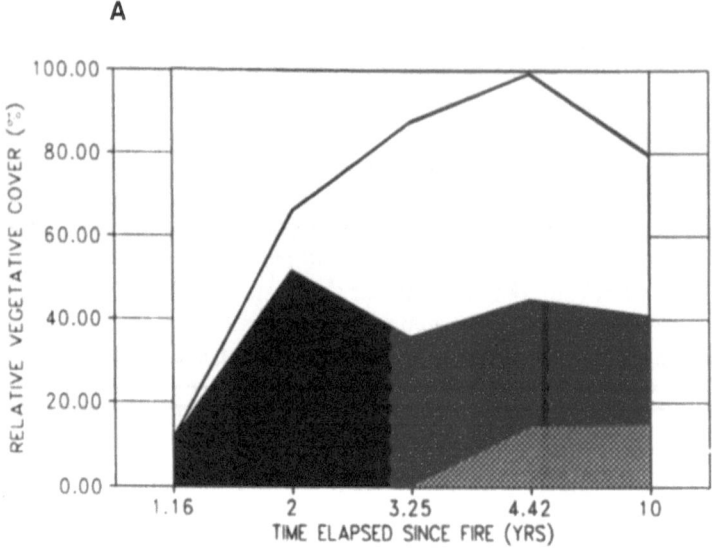

B

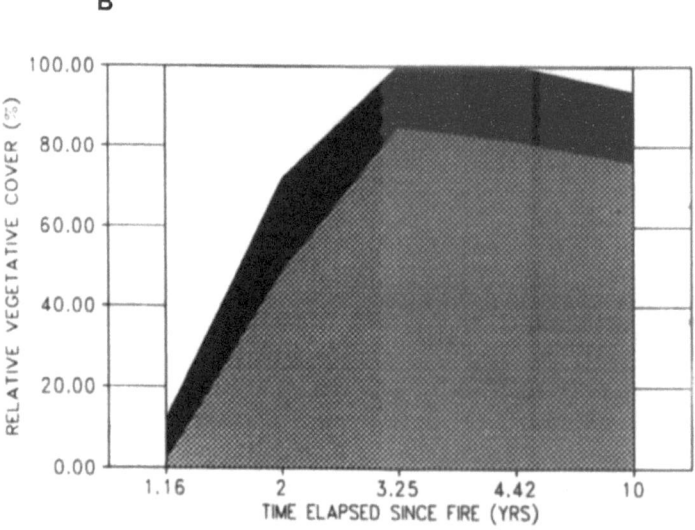

Figure 4–7. Relative vegetative cover of shrubs for 10 years following chaparral fire, with and without vertebrate herbivory. (**A**) Panel A displays, from bottom to top, relative vegetative cover with vertebrate herbivory for *Ceanothus crassifolius* (crosshatched), other species of chaparral shrubs (diagonal shading), *Salvia mellifera* (black), and herbaceous (upper left) and suffrutescent (upper right) plants. (**B**) Panel B displays relative vegetative cover of the same plant groups in the absence of herbivory by vertebrates larger than rodents.

(*Sylvilagus* sp.), and a 50-kg artiodactyl (*Odocoileus hemionus*). There were no species of nonnative mammalian herbivores present. Shifts from *Ceanothus* chaparral to coastal sage scrub after fire have been attributed to the fire regime and the presence of species of coastal sage plants (Malanson and O'Leary 1985), and to the fire regime combined with interspecific competition for light (Gray 1983). These experimental results, based on excluding herbivores, implicate plant predators as another potential cause. Mills (1983, 1986) stressed that herbivores possess the ability to modify the outcome of interspecific competition between young shrubs.

Herbivory may not always be the only cause, or by itself sufficient cause, for shifts in the plant species composition of chaparral after fire. The increased dominance of *C. crassifolius* over *A. fasciculatum* after fire has also been attributed to the greater drought tolerance of the *Ceanothus* seedlings during an unusually dry growing season following fire (Jacks 1984). Greater tolerance to summer drought has since been demonstrated for *Ceanothus* seedlings, as compared to seedlings of *A. fasciculatum* (Frazer and Davis 1988). Other variables such as postfire drought aside, experimental exclusion of herbivores after fire has clearly shown that the outcome of postfire regeneration of sclerophyll shrublands of California can be quite different in the absence of herbivory.

There is evidence that other Mediterranean-type ecosystems are also strongly shaped by herbivory. Vertebrate exclosures after fire in dry sclerophyll woodland of southeastern Australia affected the survivorship, regeneration, and growth rate of understory plants (Leigh and Holgate 1979). Both plant species composition and species densities were affected. At one site the common understory shrub *Davesia mimosoides* was much larger, denser, and had higher survivorship when it was protected from herbivory by exclosures after a small prescribed fire (Figure 4–8). Postfire exclosures against macropods in the understory of sclerophyll forests of southwestern Australia resulted in higher plant biomass and altered plant species composition within the fences (P. Christensen, personal communication).

The potential impact of herbivores on the vegetation of Mediterranean-type ecosystems has been further demonstrated in places where grazing and browsing animals are present in unusually high densities, or introduced from elsewhere. Islands provide excellent examples of these circumstances. Rottnest Island, off the coast of southwestern Australia, has an unusually dense population of the 3-kg macropod *Setonix brachyurus* (quokka), because of lack of predators. After a wildfire, these herbivores were prevented by fences from feeding in some areas. Outside the fences this led to a dramatic expansion of heathland vegetation, while within the exclosures the low forests present before the fire regenerated (Hesp et al. 1983). Similarly, a large population of the 6-kg macropod *Macropus eugenii* (Tammar Wallaby) lives on Garden Island, also near the coast of

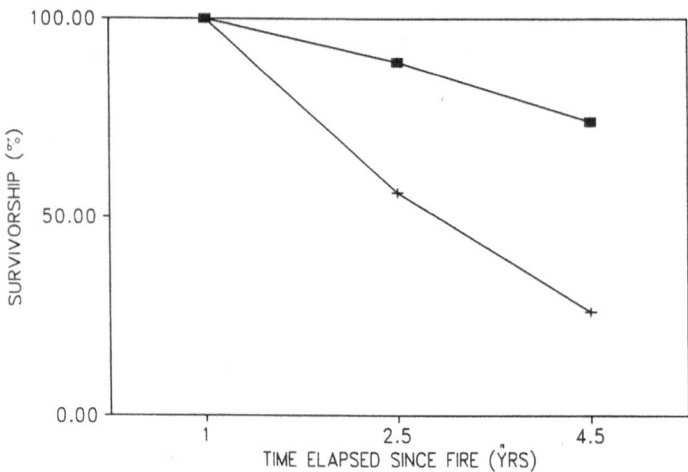

Figure 4–8. Survivorship of *Davesia mimosoides*, with (lower curve) and without (upper curve) grazing by vertebrates, for 4.2 years after fire. (Data from Leigh and Holgate [1979].)

southwestern Australia. After fire the plant community that developed inside $100\,m^2$ exclosures against native mammals was distinctly different than the plant community outside (Bell et al. 1987). Some species of plants heavily grazed outside the fences were quite common inside, and other species absent inside were present outside the exclosures. Presumably, survivorship and the outcome of interspecific plant competition were affected by the feeding preferences of the tammars. For many years, large numbers of feral goats (*Capra hircus*), pigs (*Sus scrofa*), and sheep (*Ovis aires*) have been present in various combinations on the Channel Islands, off the coast of southern California. These exotic species, introduced a century or more ago, have caused marked changes in the islands' vegetation, reducing the extent of both shrublands and woodlands through heavy browsing, grazing, and soil disturbance (Minnich 1980, 1982). Thirteen species of Channel Island plants are classified as endangered by the State of California as a direct or indirect result of herbivory by these exotic mammals (Anonymous 1990; Steinhart 1990). The islands also have two species of birds, one mammal, and one reptile that have become rare or endangered because they depend on resources provided by the diminished native plant communities.

Herbivory and the Chaparral Understory

Mature chaparral almost entirely lacks an understory beneath the 1- to 4-m high shrubs (Hanes 1981), even though seeds are present in the soil in abundance (Zammit and Zedler 1988), and manifest themselves as a dense growth of seedlings that appears only in the first or second growing seasons after fire (Keeley et al. 1981). The absence of an understory has been attributed to several causes, including: (1) competition for limiting resources (Christensen and Muller 1975a); (2) phytotoxins that inhibit germination and growth (McPherson and Muller 1969; Christensen and Muller 1975b); and (3) herbivory (Mooney and Dunn 1970; Christensen and Muller 1975a). When vertebrate herbivores have been experimentally excluded from the soil surface beneath chaparral, seedlings of herbaceous and perennial plants appeared during the winter–spring growing season (McPherson and Muller 1969; Christensen and Muller 1975a). However, in the absence of other manipulations most of these protected seedlings died at or before the end of the first growing season. These and other studies using small exclosures ($<2\,m^2$) against vertebrates have found understory seedling densities during the growing season of <1 to $71\,m^{-2}$, far less than densities reported when the shrub overstory is removed by clearing or fire (McPherson and Muller 1969; Christensen and Muller 1975b; Larson 1985).

When I placed $1\,m^2$ vertebrate exclosures beneath mature *C. crassifolius* chaparral, 11 species of plants grew in combined densities of 9 to $15\,m^{-2}$. Unlike the other studies, some plants appeared or persisted in midsummer, when moisture from precipitation was absent from the upper layer of soil. Many of these understory plants were quite small and would have gone unnoticed without close inspection. Adjacent controls, open to vertebrates, had almost no small plants in any season, and none that persisted. Distribution of herbaceous plants was quite uneven within and between exclosures. Seedlings were most frequent at microsites where litter was absent, and where small depressions and cracks presumably contained concentrations of seeds. Microtopography has been previously observed to be important in the distribution of seeds and seedlings in chaparral (Swanck and Oechel 1991). Annual grasses were by far the most numerous plants, but they never grew with sufficient vigor or survived long enough to mature and produce seeds. Within the exclosures a few perennial plants grew in densities of 1 to $4\,m^{-2}$. *Ribes* sp. (Saxifragaceae), *Galium angustifolium* Nutt. in T & G. (Rubiaceae), and the geophyte *Chlorgalum pomeridianum* (DC.) Kunth. (Hyacinthaceae) grew large enough in one growing season to produce substantial cover (50 to 100% in the $1 \times 1\,m$ plots), and to emerge through the wire mesh tops of the 30-cm-high exclosures. These plants survived and continued to grow seasonally for 3 years, escaping the protective bounds of the exclosures. These results show that seedling germination and growth was not always confined to months when surface moisture was supplied by recent rainfall, and that in the absence of vertebrate herbivory some perennial plants could become established in the understory. In the presence of herbivores, all germinating plants were soon consumed. In contrast, perennial plants are sometimes established from seedlings without protection from herbivory, in chaparral beneath the canopy (Keeley 1986b; Parker and Kelly 1989), and in canopy gaps (Keeley 1986a). How these young plants escaped destruction by herbivores is not known.

The consumption of herbaceous plants, seedlings, and resprouts of perennial shrubs in chaparral understories, clearings, and recent burns has been attributed to rabbits only (Swanck and Oechel 1991; Mills 1983), rabbits and rodents (McPherson and Muller 1969; Bartholomew 1970; Christensen and Muller 1975b), and mule deer (*O. hemionus*) (Frazer and Davis 1988). Each of these classes of herbivores is abundant and widespread in chaparral (Quinn 1990), and when present undoubtedly contributes to the commonly observed consumption of young vegetation.

The effects of various kinds of herbivores have been separated by constructing exclosures that allow access to some size classes of vertebrate consumers, while excluding others. Sets of small ($0.21\,m^2$) exclosures established in burned chaparral, and in chaparral clearings constructed by removing all shrubs, showed significant increases in the density of seedlings and plant cover within exclosures against all vertebrates, as compared to exclosures for rabbits only, and controls (Larson 1985). Plant cover in the vertebrate exclosures increased from 0 to 100% during one growing season of 4 months. Only granivory and herbivory by rodents produced a significant reduction in seedling density and cover, even though rabbits (*Sylvilagus* sp.) were present in the area and probably consumed seedlings. There was abundant evidence in the experimental area of the presence and foraging activities of rodents. Exclusion of rabbits only was insufficient to produce plant growth significantly greater than the controls.

As mature chaparral ages, deaths of individual shrubs often leave openings that develop very little new vegetation (Christensen and Muller 1975a). The absence of plants in these gaps in chaparral dominated by *A. fasciculatum* is caused by herbivory, competition with the root systems of adjacent shrubs, and lack of sufficient water and nutrients (Swanck and Oechel 1991). Among these causes, herbivory has been found to be the most important; very little herbaceous growth occurred after experimental manipulations involving interspecific competition for limiting resources, unless vertebrate herbivores were excluded (Swanck and Oechel 1991). Seven herbaceous species did survive to maturity in unmanipulated plots beneath the shrub canopy, and some of these herbs were armed with structures that could be mechanical defenses against herbivores.

Herbivory in gaps may be related to gap size. To the extent that gaps produce greater densities of young plants than the understory of the surrounding shrubs, herbivory there is probably more intense (Quinn 1986). Small areas with relatively dense growth of young plants provide a concentrated food patch for herbivores, potentially attracting them from the surrounding plant community. There is experimental evidence for concentrated herbivory in small patches in chaparral (Howe 1981; Taber and Dasmann 1958), and in the Mediterranean-type ecosystems of Australia and South Africa (Quinn 1986).

In summary, herbaceous and perennial plants do sometimes grow in gaps and beneath the canopy of mature chaparral, but the germination and growth of most of these plants is limited by a number of factors, and omnipresent herbivores consume almost all plants that are not protected.

Herbivory, Shrubland Boundaries, and Bare Zones

Where shrublands and grasslands occur adjacent to one another in the Mediterranean climate areas of California, a narrow zone, lacking either grasses or shrub cover, often occurs at the ecotone between the two communities (Bartholomew 1970; Halligan 1973; Larson 1985). The cause of this bare zone has variously been attributed to allelopathic inhibition by the shrubs (Muller et al. 1964), cattle grazing (Wells 1964), grazing by small mammals (Bartholomew 1970), and a combination of herbivory by small mammals plus allelopathy (Halligan 1973). These bare zones quickly disappeared when small mammals were experimentally excluded, evidence that herbivory and granivory alone were sufficient to explain the absence of vegetation (Bartholomew 1970; Halligan 1973; Larson 1985). Rabbits (*Sylvilagus* sp.), and some species of rodents, tend to concentrate their feeding activities in these zones, venturing only far enough away from the protective cover of shrubs to obtain food (Bartholomew 1970; Bradford 1976; Larson 1985). In this sense, the bare zone of chaparral can be viewed as a horizontal extension of the chaparral understory into an adjacent plant community – in both places small plants are infrequent due to vertebrate herbivory.

Herbivory and Chaparral Shrubs

After fire sweeps through mature chaparral, most species of burned shrubs resprout from surviving root systems (Keeley 1977, 1986b). These resprouting branches, when young and tender, are susceptible to herbivory by vertebrates, even among species that have unpalatable foliage and stems when mature (Taber and Dasmann 1958; Davis 1967; Moreno and Oechel 1991b). After fire, it is common to see the resprouting branches of many species of shrubs browsed back almost to the base of the plant. Repeated browsing of sprouts of the chaparral shrub *A. fasciculatum* after fire can significantly reduce the size (Davis 1967) and survivorship of resprouting individuals (Moreno and Oechel 1991b). Selective exclosures showed that the ungulate *O. hemionus* (Mule Deer), and not *Sylvalagus* sp. (rabbits) were responsible for growth retardation of the resprouting *A. fasciculatum* (Davis 1967). Mortality from browsing was highest among shrubs that sprouted out of synchrony with the surrounding population, suggesting that herbivory was more concentrated on such plants because less of the same food was available nearby (Moreno and Oechel 1991b). Herbivores may have been a source of selection for the usual pattern of synchronous resprouting by populations of burned *A. fasciculatum*, since this would lessen the impact of browsing on each individual by distributing it among a larger number of plants.

As described earlier, when vertebrate herbivores larger than rodents were excluded from a 10 m × 15 m plot of *C. crassifolius* chaparral for 10 years after fire, *Ceanothus* reestablished itself as the dominant species (see Figure 4–7B). In the surrounding control area, *Ceanothus* grew in much lower densities (see Figure 4–7A), apparently due to granivory in the months following the fire (see Figure 4–6), and herbivory in the decade that followed. After the first year of growth, unprotected *Ceanothus* plants were conspicuously browsed by the ungulate *O. hemoinus* (Mule Deer). These animals ate the buds and new shoots of *Ceanothus* occurring below a height of approximately 1.5 m. Feeding by mule deer retarded the growth and reproduction of all unprotected *Ceanothus* shrubs in the 1-ha study area. Browsed shrubs developed a hedged appearance because all new growth projecting away from the center of each plant was bitten back to form a smooth outside surface of short, woody, mature branches (see Figure 4–3, Stage VI). Not until the eighth year after fire did this heavily browsed population of shrubs begin to escape from complete hedging. Escape occurred when branches near the center of the hemispherical shrubs, the most difficult to reach as the shrubs gradually grew larger, began to grow upward, above the reach of the deer. A decade after the fire, unprotected plants were still significantly shorter than *Ceanothus* within the exclosure. There is qualitative evidence that the same hedging phenomenon occurred 21 years earlier, after a previous fire. Mature *C. crassifolius* shrubs in an adjacent control area, which did not burn in the most recent fire, possessed a shadow of their former hedged physiognomy, a dense hemispherical cluster of short dead branches at the bases of living shrubs. The shape of these arrays of dead branches resembled that of the young *Ceanothus* nearby, which had been hedged since the most recent fire. The hedging phenomenon may have been more noticeable at this study site than at most other experimental areas in Californian chaparral, because of a larger deer population. The study site was located within the San Dimas Experimental Forest, a US Forest Service research area and a Man and the Biosphere reserve in California that is closed to general public access and deer hunting. Deer were frequently seen in and around the study site. Most other areas of Californian chaparral are open to seasonal deer hunting. In this respect, the intense browsing here by deer may have more closely resembled the ecosystem process as it was before human intervention than do other areas of Californian chaparral with fewer deer.

These results may apply to chaparral anywhere deer are abundant. After fire, within exclosures against mule deer and free ranging cattle, the obligate seeding shrub *Ceanothus cuneatus* grew much faster than unprotected individuals (Gibbens and Schultz 1962). *C. cuneatus* outside the exclosure were hedged by deer. *Ceanothus leucodermis*, a resprouter, although browsed outside the exclosures, grew relatively rapidly and was less affected by herbivory than *C. cuneatus*. The effects of browsing were so pronounced on *C. cuneatus* and certain other shrubby species that it was suggested that deer browsing be used as a technique for controlling the growth and reproduction of chaparral shrubs in areas managed for cattle grazing (Gibbens and Schultz 1962).

Conclusion

There is not a common response pattern to fire for all species of vertebrates in Mediterranean-type ecosystems (see Table 4–1). There are differences among species for a single Mediterranean-type ecosystem, as well as differences between continents. A higher level of similarity would require convergent evolution in community functions for areas with very different biogeographical histories. Of all classes of animals, the most postfire data are available for small mammals. At the population level, there are some common patterns among small mammals. When the properties of postfire communities of small mammals after fire were compared between California, Australia, and South Africa, it was concluded that California and South Africa were similar, but Australia was different (Fox et al. 1986). Australian patterns may fail to fit the others due to a different species pool, differences in the Australian fire regime, or any number of other variables. Among birds, there are fewer studies and less quantitative data. This limits the ability to search for common patterns. Bird communities of the Mediterranean-type communities studied do seem less affected by fire than mammals. If anything, fire slightly increased species richness and overall abundance in the study areas. However, one study of the avian community of Californian coastal sage scrub did show a decrease in species richness after fire, a pattern quite different than that of nearby chaparral and other Mediterranean-type ecosystems (Stanton 1986). Data for reptiles are simply too limited to attempt intercontinental generalizations (see Table 4–1). Invertebrates may have the potential to reveal more about responses of animals to fire than any of the other taxa, because there are so many individuals and species to measure and to compare. For example, the apparent similarity between the effects of fire on soil and litter dwelling invertebrates in Australia and France, and the difference between these and the foliage dwellers in California may be generally true, but it will require more studies to permit definitive comparisons within and between ecosystems.

Available evidence suggests that herbivory by vertebrates is a significant force shaping the structure of Mediterranean-type ecosystems, and the reproduction of some species of plants. Shrubs of the genus *Ceanothus* provide a case study of the pervasive effects of vertebrate herbivory. Of the six stages in the life history of *C. crassifolius*, all but one receive significant impacts from plant predators (see Figure 4–1). Rodents deplete the soil seed bank of this species and probably consume seedlings, whereas rabbits consume seedlings and young plants in preference to competing plant species, and deer exert browsing pressure sufficient to retard growth and reproduction for at least a decade after fire.

Vertebrate exclosures have proven to be useful tools for identifying and quantifying impacts of vertebrate herbivory on plant communities as they regenerate after fire. Small vertebrate exclosures within and at the edges of Californian shrublands have provided convincing evidence that vertebrates are the most important variable limiting the development of a plant community beneath and adjacent to shrubs, and in gaps between shrubs. Exclosures are strongly recommended as a useful and efficient way for gaining insight into connections between herbivory and plant community development after disturbance.

The conclusions about vertebrate herbivory, based for the most part on studies from Californian chaparral, support the viewpoint that animals are a powerful element shaping plant communities, especially as they regenerate after fire. Vertebrates may play similar roles in other fire-prone Mediterranean-type ecosystems, but conclusions and comparisons must await further field work.

Acknowledgment. I thank Barbara Ellis-Quinn for help in preparation of the manuscript, V. Rosales for assistance with the figures, and Tony Cario for data analysis. Much of the field work on *Ceanothus* was done by students, and I am grateful to them all. Miguel Angel de Zavala contributed time to the project for an entire summer. The support and encouragement of the personnel of the San Dimas Experimental Forest, in particular C. Colver, S. Conard, C. Conrad, P. Dunn, and D. Larson was indispensable. Field work was supported by Pacific Southwest Forest and Range Experiment Station, USDA Forest Service, the Cal Poly Kellogg Unit Foundation, and California State Polytechnic University, Pomona, CA.

References

Abbott, I. 1984. Changes in the abundance and activity of certain soil and litter fauna in the jarrah forest of Western Australia after a moderate intensity fire. *Australian Journal of Soil Research* 22(4):463–470.

Anonymous, 1990. 1989 *Annual Report on the Status of California's State Listed Threatened and Endangered Plants and Animals.* State of California, Department of Fish and Game. Sacramento, CA.

Athias-Binche, F. 1987. Modalités de la cicatrisation des écosystèmes Méditerranéens après incendie: Cas de certains arthropodes du sol. 3. les acariens Uropodides. Vie Milieu 37:39–52.

Auld, T. 1986. Population dynamics of the shrub *Acacia suaveolens* (Sm.) Willd.: Fire and the transition to seedlings. *Australian Journal of Ecology* 11:373–385.

Bamford, M. 1985. The fire-related dynamics of small vertebrates in *Banksia* woodland: A summary of research in progress, pp. 107–110. *In* J. Ford (ed.), *Symposium on fire ecology and management in Western Australian Ecosystems*. W.A.I.T. Env. Studies Group Pub. No. 14. Western Australia Institute of Technology, Western Australia.

Barro, S., Quinn, R. 1991. *Ceanothus* seed viability over time and its relevance to seed pool dynamics. Unpublished manuscript.

Bartholomew, B. 1970. Bare zone between California shrub and grassland communities: The role of animals. *Science* 170:1210–1212.

Bayless, C. 1980. Microhabitats in a chaparral rodent community. Unpublished Master's thesis, California State Polytechnic University, Pomona, CA.

Bell, B., Moredoundt, J., Loueragan, W. 1987. Grazing pressure by the tammar (*Macropus eugenii* Desm.) on the vegetation of Garden Island, Western Australia, and the potential impact on food reserves of a controlled burning regime. *Journal of the Royal Society of Western Australia* 69:89–94.

Biswell, H. 1961. Manipulation of chamise brush for deer range improvement. *California Fish and Game* 47:125–144.

Biswell, H. 1989. *Prescribed burning in California wildlands vegetation management*. University of California Press, Berkeley, CA.

Bornemizza, G. 1969. The re-invasion of burnt woodland areas by insects and mites. Paper presented at seminar on the effects of Forest Fire, Ecological Society of Australia, Canberra Group.

Bradford, D. 1976. Space utilization by rodents in *Adenostoma* chaparral. *Journal of Mammalogy* 57:576–579.

Breytenbach, G. 1987. Small mammal dynamics in relation to fire, pp. 56–68. *In* R. Cowling, D. Le Maitre, B. McKenzie, R. Prys-Jones, and B. vanWilgen (eds.), *Disturbance and the dynamics of fynbos biome communities*. South African National Scientific Programmes Report No. 135. Council for Scientific and Industrial Research, Pretoria, South Africa.

Bullock, S. 1978. Plant abundance and distribution in relation to types of seed dispersal in chaparral. *Madroño* 25:104–105.

Catling, P., Newsome, A. 1981. Responses of the Australian vertebrate fauna to fire: An evolutionary approach, pp. 273–310. *In* A. Gill, R. Groves, and I. Noble (eds.), *Fire and the Australian Biota*. Australian Academy of Science, Canberra, Australia.

Catling, P., Newsome, A., Dudzinski, G. 1982. Small mammals, habitat components, and fire in Southeastern Australia, pp. 199–206. *In* C. Conrad and W. Oechel (eds.), *Dynamics and Management of Mediterranean-type Ecosystems*. Pacific Southwest Forest and Range Experiment Station, US Forest Service, Berkeley, CA.

Chew, R., Butterworth, B., Grechman, R. 1959. The effects of fire on the small mammal populations of chaparral. *Journal of Mammalogy* 40:253.

Christensen, N., Muller, C. 1975a. Relative importance of factors controlling germination and seedling survival in *Adenostoma* chaparral. *American Midland Naturalist* 93:71–78.

Christensen, N., Muller, C. 1975b. Effects of fire on factors controlling plant growth in *Adenostoma* chaparral. *Ecological Monographs* 45:29–55.

Christensen, P., Kimber, P. 1975. Effects of prescribed burning on the flora and fauna of Southwest Australian forest. *Proceedings of the Ecological Society of Australia* 9:85–106.

Christensen, P., Recher, H., Hoare, J. 1981. Responses of open forests (dry sclerophyll forests) to fire regimes, pp. 367–393. *In* A. Gill, R. Groves, and I.

Noble (eds.), *Fire and the Australian Biota*. Australian Academy of Science, Canberra, Australia.

Countryman, C. 1974. Can southern California wildland conflagrations be stopped? *USDA Forest Service General Technical Report* PSW–7. Pacific Southwest Forest and Range Experiment Station, Berkeley, CA.

Cowley, R. 1974. Effects of prescribed burning on birds of the mixed species forests of west central Victoria, pp. 58–65. *In Proceedings of Third Fire Ecology Symposium*, Monash University, Melbourne, Australia.

Davey, J. 1982. Stand replacement in *Ceanothus crassifolius*. Unpublished Master's thesis, California State Polytechnic University. Pomona, CA.

Davis, J. 1967. Some effects of deer browsing on chamise sprouts after fire. *American Midland Naturalist* 77:234–238.

DeBano, L., Rice, M., Conrad, C. 1979. Soil heating in chaparral fires: Effects on soil properties, plant nutrients, erosion, and runoff. United States Forest Service, Research Paper PSW–145.

Force, D. 1981. Postfire insect succession in southern California chaparral. *American Naturalist* 117:575–582.

Force, D. 1982. Postburn insect fauna in southern California chaparral, pp. 234–240. *In* C. Conrad and W. Oechel (eds.), *Dynamics and Management of Mediterranean-Type Ecosystems*. Pacific Southwest Forest and Range Experiment Station, US Forest Service, Berkeley, CA.

Fox, A. 1978. The '72 fire of Nadgee Nature Reserve. *Parks and Wildlife* 2(2): 5–24.

Fox, B. 1982. Fire and mammalian secondary succession in an Australian coastal heath. *Ecology* 63.(5):1332–1341.

Fox, B., Quinn, R., Breytenbach, G. 1986. A comparison of small-mammal succession following fire in shrublands of Australia, California and South Africa. *Proceedings of the Ecological Society of Australia* 14:179–197.

Fox, B., Fox, M. 1987. The role of fire in the scleromorphic forests and shrublands of Eastern Australia, pp. 23–48. *In* L. Trabaud (ed.), *The Role of Fire in Ecological Systems*. SPB Academic Press, The Hague, The Netherlands.

Frazer, J., Davis, S. 1988. Differential survival of chaparral seedlings during the first summer drought after wildfire. *Oecologia* 76:215–221.

Gibbens, R., Schultz, A. 1962. Manipulation of shrub form and browse production in game range improvement. *California Fish and Game* 48:49–64.

Gray, J. 1983. Competition for light and a dynamic boundary between chaparral and coastal sage scrub. *Madroño* 30:43–49.

Halligan, J. 1973. Bare areas associated with shrub stands in grassland: The case of *Artemisia californica*. *Bioscience* 23:429–432.

Hanes, T. 1977. Chaparral, pp. 417–470. *In* M. Barbour and J. Major (eds.), *Terrestrial Vegetation of California*. John Wiley & Sons. New York.

Hanes, T. 1981. California chaparral, pp. 139–144. *In* F. DiCastri, D. Goodall, and R. Specht (eds.), *Ecosystems of the World, Volume 11, Mediterranean-Type Shrublands*. Elsevier Scientific, New York.

Herrera, C. 1984. A study of avian frugivores, bird-dispersed plants, and their interaction in Mediterranean scrublands. *Ecological Monographs* 54:1–23.

Hesp, P., Wells, M., Ward, B., Riches, J. 1983. *Land resource survey of Rottnest Island*. Bulletin No. 4086, Soil and Conservation Service Branch, Division of Resource Management, Department of Agriculture, Western Australia.

Horton, J., Kraebel, C. 1955. Development of vegetation after fire in the chamise chaparral of southern California. *Ecology* 36(2):244–262.

Howard, W., Fenner, R., Childs, H. 1959. Wildlife survival in brush burns. *Journal of Range Management* 12:230–234.

Howe, G. 1981. Death of chamise (*Adenostoma fasciculatum*) shrubs after fire or cutting as a result of herbivore browsing. *Bulletin of Southern California Academy of Sciences* 80:138–143.

Howe, G., Carothers, L. 1980. Postfire seedling reproduction of *Adenostoma fasciculatum* H. and A. *Bulletin of Southern California Academy of Sciences* 79(1):5–13.

Jacks, P. 1984. The drought tolerance of *Adenostoma fasciculatum and Ceanothus crassifolius* seedlings and vegetation change in the San Gabriel chaparral. Unpublished Master's thesis, San Diego State University, San Diego, CA.

Kahn, W. 1960. Observations on the effect of a burn on a population of *Sceloporus occidentalis*. *Ecology* 41:358–359.

Keeley, J. 1977. Fire-dependent reproductive strategies in *Arctostaphylos* and *Ceanothus*, pp. 391–396. *In* H. Mooney and C. Conrad (eds.), *Symposium on the Environmental Consequences of Fire and Fuel Management in Mediterranean Ecosystems*. USDA Forest Service General Technical Report WO–3. USDA Forest Service, Washington, DC.

Keeley, J. 1986a. Seed germination patterns of *Salvia mellifera* in fire-prone environments. *Oecologia* 71:1–5.

Keeley, J. 1986b. Resilience of Mediterranean shrub communities to fires, pp. 391–396. *In* B. Dell, A. Hopkins, and B. Lamont (eds.), *Resilience of Mediterranean-Type Ecosystems*. Dr. W. Junk, Dordrecht, The Netherlands.

Keeley, J. 1987. Ten years of change in seed banks of the chaparral shrubs, *Arctostaphylos glauca* and *Arctostaphylos glandulosa*. *American Midland Naturalist* 117:446–448.

Keeley, J., Hays, R. 1976. Differential seed predation on two species of *Arctostaphylos* (Ericaceae). *Oecologia* 24:71–81.

Keeley, S., Keeley, J., Hutchinson, S., Johnson, A. 1981. Postfire succession of the herbaceous flora in southern California chaparral. *Ecology* 62(6): 1608–1619.

Kimber, P. 1974. Some effects of prescribed burning on jarrah forest birds, pp. 49–57. *Proceedings of the Third Fire Ecology Symposium*. Monash University, Melbourne, Australia.

Kruger, F., Bigalke, R. 1984. Fire in fynbos, pp. 67–114. *In* P. de V. Booysen and N. Tainton (eds.), *Ecological Effects of Fire in South African Ecosystems*. Springer-Verlag, New York.

Larson, D. 1985. Habitat utilization, diet, and herbivory effects of rabbits in southern California chaparral. Unpublished Master's thesis, California State Polytechnic University. Pomona, CA.

Lawrence, G. 1966. Ecology of vertebrate animals in relation to chaparral fire in the Sierra Nevada foothills. *Ecology* 47(2):278–291.

Leigh, J., Holgate, W. 1979. The response of the understorey of forests and woodlands of the Southern Tablelands to grazing and browsing. *Australian Journal of Ecology* 4:23–43.

Leonard, B. 1974. Effects of burning on litter fauna in eucalypt forest, pp. 42–46. *Proceedings of the Third Fire Ecology Symposium*. Monash University, Melbourne, Australia.

Lillywhite, H. 1977a. Effects of chaparral conversion on small vertebrates in southern California. *Biological Conservation* 11:171–184.

Lillywhite, H. 1977b. Animal responses to fire and fuel management in chaparral, pp. 368–373. *In* H. Mooney and C. Conrad (eds.), *Proceedings of the Symposium on the Environmental Consequences of Fire and Fuel Management in Mediterranean Ecosystems*. Pacific Southwest Forest and Range Experiment Station, USDA Forest Service General Technical Report WO–3. Berkeley, CA.

Lloret, F., Zedler, P. 1991. Recruitment pattern of *Rhus integrifolia* populations in periods between fire in chaparral. *Journal of Vegetation Science* 2:217–230.

Longhurst, W. 1978. Responses of bird and mammal populations to fire in chaparral. *California Agriculture* 10(10):9–12.

Lunney, D., Ashby, E., Grigg, J., O'Connell, M. 1989. Diets of scincid lizards *Lampropholis guichenoti* (Dumeril & Bibron) and *L. delicata* (De Vis) in Mumbulla State Forest on the South Coast of New South Wales. *Australian Wildlife Research* 16:307–312.

Lunney, D., Barker, J. 1986. Survey of reptiles and amphibians of the coastal forests near Bega, NSW. *Australian Zoologist* 22(3):1–9.

Majer, J. 1985. Fire effects on invertebrate fauna of forest and woodland, pp. 103–106. *In* J. Ford (ed.), *Symposium on fire ecology and management in Western Australian Ecosystems*. W.A.I.T. Environmental Studies Group Pub. No. 14. Western Australia Institute of Technology, Western Australia.

Malanson, G., O'Leary, J. 1985. Effects of fire and habitat on post-fire regeneration in Mediterranean-type ecosystems: *Ceanothus spinosus* chaparral and Californian coastal sage scrub. *Acta Oecologia/Oecologia Plantarum* 20 (new series volume 6):169–181.

McPherson, J., Muller, C. 1969. Allelopathic effects of *Adenostoma fasciculatum*, "chamise", in the California chaparral. *Ecological Monographs* 39:178–197.

Mills, J. 1983. Herbivory and seedling establishment in post-fire southern California chaparral. *Oecologia* 60:267–270.

Mills, J. 1986. Herbivores and early postfire succession in southern California chaparral. *Ecology* 67:1637–1649.

Minnich, R. 1980. Vegetation of Santa Cruz and Santa Catalina Islands, pp. 123–138. *In* D. Power (ed.), *The California Islands*. Santa Barbara Museum of Natural History. Santa Barbara, California.

Minnich, R. 1982. Grazing, fire, and the management of vegetation on Santa Catalina Island, California, pp. 444–449. *In* C. Conrad and W. Oechel (eds.), *Dynamics and Management of Mediterranean-Type Ecosystems*. Pacific Southwest Forest and Range Experiment Station. Berkeley, CA.

Minnich, R. 1989. Chaparral fire history in San Diego County and adjacent northern Baja California: An evaluation of natural fire regimes and the effects of suppression management, pp. 37–47. *In* S. Keeley (ed.), *The California Chaparral: Paradigms Reexamined*. No. 34 Science Series, Natural History Museum of Los Angeles County, Los Angeles, CA.

Moldenke, A. 1977. Insect-plant relationships, pp. 199–217. *In* N. Thrower and D. Bradbury (eds.), *Chile-California Mediterranean Scrub Atlas*. Dowden, Hutchinson, & Ross, Stroudsburg, PA.

Mooney, H., Bonnicksen, T., Christensen, N., Lotan, J., Reiners, W. (eds.) 1981. *Proceedings of the Conference on Fire Regimes and Ecosystem Properties*. USDA, Forest Service, General Technical Report WO-3.

Mooney, H., Dunn, E. 1970. Convergent evolution of Mediterranean-climate evergreen sclerophyll shrubs. *Evolution* 24:292–303.

Moreno, J.M., Oechel, W.C. 1991a. Fire intensity effects on the germination of shrub and herbaceous species in southern California chaparral. *Ecology* 72:1993–2004.

Moreno, J.M., Oechel, W.C. 1991b. Fire intensity and herbivory effects on postfire resprouting of *Adenostoma fasciculatum* in southern California chaparral. *Oecologia* 85:429–433.

Muller, C., Muller, W., Haines, B. 1964. Volatile growth inhibitors produced by aromatic shrubs. *Science* 143:471–473.

Parker, V., Kelly, V. 1989. Seed banks in California chaparral and other Mediterranean climate shrublands, pp. 231–255. *In* M. Leck, V. Parker, and R. Simpson (eds.), *Ecology of Soil Seed Banks*. Academic Press, New York.

Price, M., Reichman, O. 1987. Distribution of seeds in Sonoran desert soils: Implications for heteromyid rodent foraging. *Ecology* 68:1797–1811.

Price, M., Waser, N. 1984. On the relative abundance of species: Postfire changes in a coastal sage scrub rodent community. *Ecology* 65(4):1161–1169.

Prodon, R., Fons R., Athias-Binche, F. 1987. The impact of fire on animal communities in Mediterranean area, pp. 121–157. *In* L. Trabaud (ed.), *The Role of Fire in Ecological Systems, 4th Congress of Ecology, Syracuse, NY*. SPB Academic Publishing, The Hague, The Netherlands.

Quick, C. 1956. Viable seeds from the duff and soil of sugar pine forests. *Forest Science* 2:36–42.

Quick, C., Quick, A. 1961. Germination of *Ceanothus* seeds. *Madroño* 16:23–30.

Quinn, R. 1979. Effects of fire on small mammals in the chaparral. *Cal-Neva Wildlife Trans.* 1979:125–133.

Quinn, R. 1983. Short-term effects of habitat management on small vertebrates in chaparral. *Cal–Neva Wildlife Transactions* 1983:55–66.

Quinn, R. 1986. Mammalian herbivory and resilience in Mediterranean-climate ecosystems, pp. 113–128. *In* B. Dell, A. Hopkins, and B. Lamont (eds.), *Resilience of Mediterranean-Type Ecosystems*. DR W Junk, Dordrecht, The Netherlands.

Quinn, R. 1990. Habitat preferences and distribution of mammals. *In California chaparral*. USDA Forest Service Research Paper PSW–202. Berkeley, CA.

Recher, H., Christensen, P. 1981. Fire and the evolution of the Australian biota pp. 137–162. *In* A. Keast (ed.), Ecological Biogeography of Australia. Dr. W. Junk, The Hague, The Netherlands.

Riggan, P., Goode, S., Jacks, P., Lockwood, R. 1988. Interaction of fire and community development in chaparral of southern California. *Ecological Monographs* 58:155–176.

Saulnier, L., Athias-Binche, F. 1986. Modalités de la cicatrisation des écosystèmes Méditerranéens après incendie: cas de certains arthropodes du sol. 2. Les Myriapodes édaphiques. *Vie Milieu* 36(3):191–204.

Schlettwein, C., Giliomee, J. 1987. Comparison of insect biomass and community structure between fynbos sites of different ages after fire with particular reference to ants, leafhoppers and grasshoppers. *Annals of the University of Stellenbosch Ser. A 3 (Landbouwet)* 2(2):1–76.

Simovich, M. 1979. Post-fire reptile succession. *Cal–Neva Wildlife Transactions* 1979:104–113.

Smith, P. 1989. Changes in a forest bird community during a period of fire and drought near Bega, New South Wales. *Australian Journal of Ecology* 14:41–54.

Springett, J. 1976. The effect of prescribed burning on the soil fauna and on litter decomposition in Western Australian forests. *Australian Journal of Ecology* 1:77–82.

Springett, J. 1979. The effects of a single hot summer fire on soil fauna and on litter decomposition in jarrah (*Eucalyptus marginata*) forest in Western Australia. *Australian Journal of Ecology* 4:279–291.

Stanton, P. 1986. Comparison of avian community dynamics of burned and unburned coastal sage scrub. *The Condor* 88:285–289.

Steinhart, P. 1990. *California's Wild Heritage; Threatened and Endangered Animals in the Golden State*. California Department of Fish and Game, Sacramento, CA.

Stubbs, D., Swingland, I., Hailey, A. 1985. The ecology of the Mediterranean Tortoise *Testudo hermanni* in Northern Greece (The effects of a catastrophe on population structure and density). *Biological Conservation* 31:125–152.

Swanck, S., Oechel, W. 1991. Interactions among the effects of herbivory, competition, and resource limitation on chaparral herbs. *Ecology* 72:104–115.

Taber, R., Dasmann, R. 1958. The black-tailed deer of the chaparral: Its life history and management in the North Coast Range of California. California Department of Fish and Game Bulletin 8. Sacramento, CA.

Wells, P. 1964. Antibiosis as a factor in vegetation patterns. *Science* 144:889.

Wells, P. 1969. The relation between mode of reproduction and extent of speciation in woody genera of the California chaparral. *Evolution* 23:264–267.

Westman, W. 1982. Coastal sage succession, pp. 91–99. *In* C. Conrad and W. Oechel (eds.), *Proceedings of the Symposium on Dynamics and Management of Mediterranean-Type Ecosystems*. USDA Forest Service Gen. Tech. Rep. PSW–58. Berkeley, CA.

Wilan, K., Bigalke, R. 1982. The effects of fire regime on small mammals in S. W. Cape montane fynbos (Cape macchia), pp. 207–212. *In* C. Conrad and W. Oechel (eds.), *Proceedings of the Symposium on Dynamics and Management of Mediterranean-Type Ecosystems*. USDA Forest Service Gen. Tech. Rep. PSW–58.

Wirtz, W. 1979. Who burns? pp. 36–37. *Abstracts of the 60th Annual Meeting of the Western Society of Naturalists*.

Wirtz, W. 1982. Postfire community structure of birds and rodents in southern California chaparral, pp. 241–246. *In* C. Conrad and W. Oechel (eds.), *Proceedings of the Symposium on Dynamics and Management of Mediterranean-Type Ecosystems*. USDA Forest Service Gen. Tech. Rep. PSW–58.

Wirtz, W., Hoekman, D., Muhm, J., Souza, S. 1988. Postfire rodent succession following prescribed fire in southern California chaparral, pp. 333–339. *In* R. Szaro, K. Severson, and D. Patton (eds.), *Management of Amphibians, Reptiles, and Small Mammals in North America*. USDA Forest Service General Technical Report RM–166.

Wooler, R., Calver, M. 1988. Changes in an assemblage of small birds in the understorey of dry sclerophyll forest in Southwestern Australia after fire. *Australian Wildlife Research* 15:331–338.

Wright, M. 1988. A note on the reaction of angulate tortoises to fire in fynbos. *South African Journal of Wildlife Research* 18:131–133.

Zammit, C., Zedler, P. 1988. The influence of dominant shrubs, fire, and time since fire on soil seed banks in mixed chaparral. *Vegetatio* 75:175–187.

5. The Effects of Fire on Physical and Chemical Properties of Soils in Mediterranean-Climate Shrublands

Norman L. Christensen

Given the rapid oxidation of above-ground biomass and the differences in prefire and postfire microclimate, it is not surprising that the biogeo-chemical consequences of fire have received considerable attention. Fire influences virtually every process illustrated in Figure 5–1; however, few generalizations about its specific effects on soils and biogeochemical process are possible. The variation in such effects is conditioned by at least four broad classes of factors: (1) variation in basic site features such as slope aspect, climate, and soil; (2) spatial and temporal variation in fire regimes; (3) prefire status of vegetation in fuels; (4) postfire patterns of ecosystem recovery. The role of these factors in determining the patterns of fire effects on soils and biogeochemical processes is a central theme of this chapter.

The Fire Regimes and Soils in Mediterranean-Climate Ecosystems

Fires in Mediterranean-climate shrublands generally burn with consider-able intensity. Average fire return times vary from 20 to 30 years to over a century, and fuel combustion during individual fire events varies from 1 to $5\,kg\,m^{-2}$ (Olson 1981). Spatial variation in fire behavior results largely from differences in prefire fuel mass, fuel moisture content, fire weather conditions, and site topography. Fire frequency (both regional frequency and return time) is regulated by the frequency and distribution of ignition

Patterns of nutrient cycling

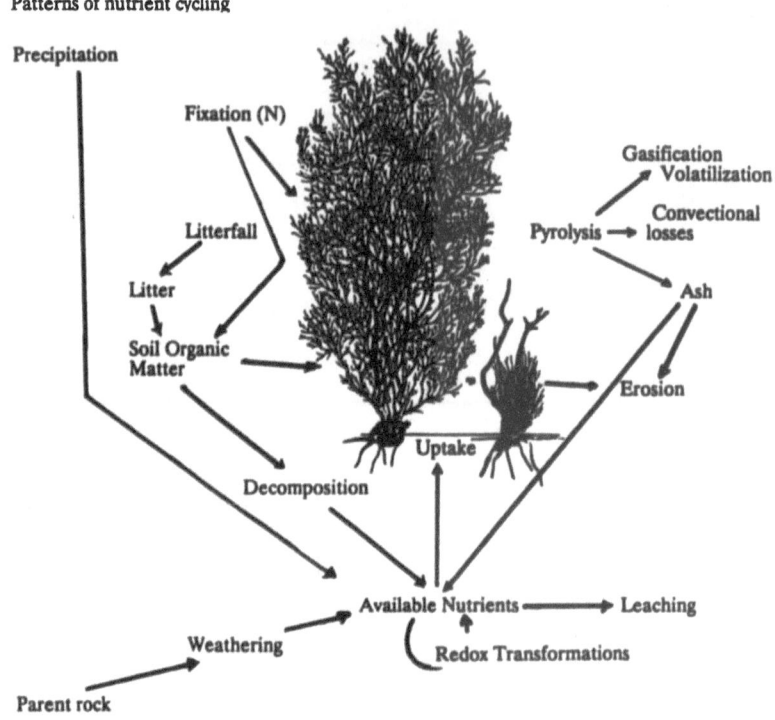

Figure 5–1. Processes affecting nutrient cycling during a fire cycle in a chaparral ecosystem. (Modified from Marion 1982.)

events, fuel accumulation rates, and landscape configuration. The significance of this last factor has received considerable attention (Minnich 1983, 1988; Turner 1987; Malanson 1984). The relative impact of topography and spatial variation in fuels on fire behavior depends on weather conditions and fuel moisture. At moderate-to-high fuel moisture, variations in vegetation structure and localized landscape fragmentation (due to past fire history) may determine burning patterns. However, when fuel moisture drops below threshold levels and weather conditions are extreme (such as hot, dry winds), fire behavior may be regulated primarily by large-scale topographic features such as major rivers or divides (Turner and Romme 1994).

There is general agreement that the above factors regulate Mediterranean-climate fire regimes, but quantitative estimates of their relative importance are difficult, if not impossible. This is due in part to limited data available, but is also a consequence of the stochastic nature of several of these factors. Thus, it is incorrect to view Mediterranean ecosystem fire regimes as being under finely tuned feedback control. Indeed, the variability in fire regimes resulting from these largely unpredictable effects may be a critical component in the overall diversity of Mediterranean-climate landscapes.

There is considerable variability in the seasonality of fires in Mediterranean-climate shrublands; fuels are driest and ignition sources are most frequent in the summer. Thus, the vast majority of fires occur in summer, and winter and early spring fires are relatively uncommon.

The impact of any particular fire regime on soils and biogeochemical processes depends on basic site characteristics such as slope, parent rock, and soil properties. Mediterranean-climates often, although not always, coincide with areas of active geologic and tectonic activity, resulting in rugged topography and steep slopes. High rates of erosion result in relatively young soils (i.e., soils with limited profile development). The ancient weathered soils of South Africa and southwestern Australia are obvious exceptions to this pattern (Boucher and Moll 1981; Specht 1981). Variations among geographic regions in leaf concentrations of nitrogen (N) and phosphorus (P) (Figure 5–2) are related to variations in the parent material and are thought to reflect variations in patterns of nutrient availability (Rundel 1979; di Castri 1981; Margaris 1982; Margaris et al. 1981). Low concentrations of N and P in plant tissue from the South African fynbos and Australian heaths are very likely a consequence of the silicaceous parent materials from which these soils were derived. Soils in these areas also tend to be quite acidic. Concentrations of phosphorus are

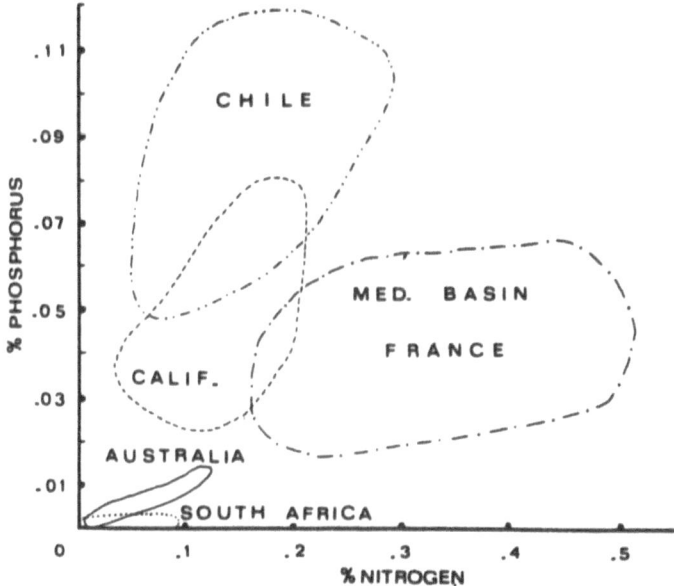

Figure 5–2. N and P content of soils of the various Mediterranean-type regions of the world. (From Rundel [1976], di Castri [1981] and Margaris et al. [1984].)

considerably higher in plant tissues from California and Chile than in tissues from other regions, and probably indicate a greater availability of this mineral from these parent rocks. The high concentrations of nitrogen in foliage sampled from the Mediterranean Basin reflect the importance of legume species in that vegetation (Quézel 1981). Although the mechanisms are not clear, these high concentrations of nitrogen are probably related to the widespread alkaline soils (Terra Rosa) derived from calcareous substrates.

The availability of both nitrogen and phosphorus often limits plant growth in Mediterranean-climate shrublands. However, the extent and relative importance of such limitations vary within and among regions. Such variations likely also affect the character and magnitude of plant responses to fire-caused changes in nutrient availability.

Physical Impacts of Fire on Soil

The extent of soil heating during fire depends on fire intensity and duration, as well as on soil water content (Figure 5–3). In general, the direct effects of intense heating are confined to the upper 2 to 3 cm of soil (DeBano et al. 1979). However, in areas with extensive fuel accumulation, smoldering fires can heat the soil profile to a depth of 10 to 20 cm, resulting in considerable chemical change and soil sterilization.

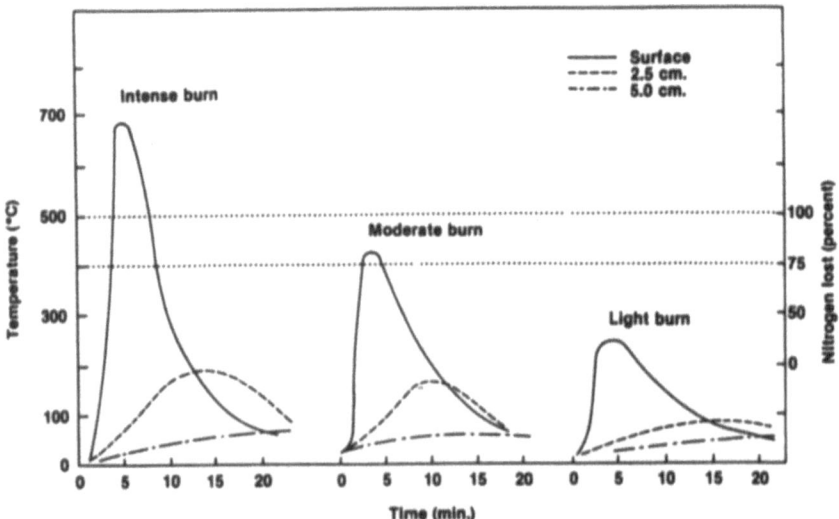

Figure 5–3. Time course of temperatures at various soil depths and associated N losses of three different types of fires in California chaparral. (From DeBano 1982.)

Except where soil heating is exteme, physical changes in soil features such as texture or mineralogy are negligible. Where the parent rock is exposed, weathering may be accelerated as a consequence of heat-caused spalling.

Fire-caused changes in the nature and distribution of organic matter may alter soil physical properties. The extent of soil organic mineral loss varies in direct proportion to soil temperature (DeBano et al. 1979). The loss of organic matter may result in decreased cation exchange capacity and increased soil bulk density. Resulting structural changes could conceivably alter soil water-holding characteristics, although such changes have not been reported.

DeBano et al. (1979) described fire-caused biochemical changes that alter hydrologic properties of some California chaparral soils. Accumulation of organic hydrophobic substances in the upper soil layers during interfire years may render the soil surface repellent to water penetration. The heat of the fire distills these hydrophobic compounds down the soil profile, creating a wettable layer of soil 2- to 3-cm thick overlaying a water-repellent soil layer (DeBano and Conrad 1978). DeBano et al. (1979) proposed that saturation of the upper wettable layer creates conditions favorable to sheet erosion. The specific compounds involved in the development of these hydrophobic soil layers have not been identified, and it is not clear whether the compounds are produced directly by plants, are products of decomposition, or are exudates from soil microbes (DeBano 1982).

Fire-Caused Nutrient Loss

Fires necessarily result in the net loss of nutrients from ecosystems, although there are surprisingly few estimates of such losses. At least five mechanisms account for these losses: (1) oxidation of compounds to a gaseous form (gasification); (2) vaporization of compounds that were solid at normal temperatures; (3) convection of ash particles in fire-generated winds; (4) leaching of ions in solution out of the soil; (5) accelerated erosion following fire (Binkley and Christensen 1991). The relative importance of these mechanisms varies for each nutrient and is a consequence of variations in fire intensity, site soil and topography, and climatic patterns.

Organic compounds that contain nitrogen and sulfur (S) are easily oxidized during fire, and the gasification losses of N and S are usually directly proportional to the loss of organic matter (Raison et al. 1985). Data from a variety of fuel types indicate that nitrogen losses to gasification during fires are approximately $6 \, kg \, Mg^{-1}$ fuel consumed (DeBano and Conrad 1978; Binkley and Christensen 1991). Based on this value, and assuming a fire return interval of 50 to 100 years for Mediterranean-climate shrublands, losses of nitrogen can be calculated on an annual basis over a fire cycle. Expressed in this manner, annual nitrogen losses in chaparral fires may vary from approximately 0.12 to 12 $kg\,ha^{-1}\,yr^{-1}$. Such nitrogen losses are of roughly the same order of magnitude as annual nitrogen inputs in bulk precipitation (Boring et al. 1988). Thus, nitrogen fixation probably plays an important role in the long-term balance of nitrogen budgets in Mediterranean-climate shrublands.

Gasification of phosphorus compounds may also occur during fires. Phosphorus in ecosystems is virtually entirely in the form of phosphate (PO_4^{3-}). However, high temperatures in fires may generate a variety of phosphorus oxides (Cotton and Wilkinson 1988). Gasification of elements besides N, P, and S appears to be relatively unimportant (Binkley and Christensen 1991).

Vaporization losses during fire occur when an element is volatilized but its oxidation state remains unchanged. For example, nitrate (NO_3^-) volatilizes at temperatures as low as 80°C (Greenwood and Earnshaw 1984), and amino acids may vaporize at temperatures below 200°C (Weast 1982), well below their combustion temperature. Metal cations such as potassium, calcium (Ca), and magnesium (Mg), may also be vaporized in small quantities, especially if they are part of organic molecules (Raison et al. 1985).

Convection losses (or additions) of nutrients occur as ash is transported out of or into burn areas by air currents. The extent of this loss depends heavily on fire behavior and accompanying weather, but can be substantial given the high concentrations of nutrients in ash. The importance of this effect is highly dependent on scale. Ash redistribution contributes to variability in fire nutrient budgets on small spatial scales within fires. However, all other factors being equal, net loss from entire burns should diminish as the area of the burn increases.

The extent of leaching losses from soil following fire is dependent on the availability of nutrients; patterns of plant uptake and retention; the absorptive properties of litter, humus and soils; and postfire patterns of precipitation and evapotranspiration. In general, the anion and cation adsorptive characteristics of soils minimize the extent of leaching losses. Grier (1975) found that the majority of cations and anions that is released from ash is retained in the upper 20 cm of soil.

Because of the high mobility of the NO_3^- anion and the fact that fire frequently accelerates rates of nitrification, leaching may result in considerable loss of soil nitrogen capital following fire (Tiedemann et al. 1978). In certain chaparral ecosystems, such losses may be equivalent to 0.5% of the total soil nitrogen capital (Longstreth and Patten 1975; DeBano and Conrad 1978). Such losses are likely to be greater where soil texture is coarse or where there has been considerable loss of soil organic matter.

Leaching losses of cations may be directly influenced by soil heating. Stark (1977) found that soil adsorption of Ca and Mg was increased while that of iron (Fe) was decreased when soils were heated to temperatures above 300°C. Fire has been found to influence the adsorptive properties of soils for phosphorus in a variety of ecosystem types (Christensen 1987).

Accelerated rates of erosion are among the most well known effects of fire in Mediterranean shrubland ecosystems. Erosion rates may increase on slopes following fire as a consequence of changes in vegetation, soil properties, hydrology, and geomorphic processes (Swanson 1981). The actual amount and duration of such changes in erosion rates vary widely among sites as a consequence of fire intensity, soil infiltration capacity, topography, climate, and patterns of vegetation recovery (Binkley and Christensen 1991). Loss of plant cover and litter and humus layers exposes the soil to increased kinetic energy of raindrops, which in turn, increases sediment movement. These effects may be ameliorated by postfire deposition of ash and surface litter (Connaughton 1935; Megahan and Moliter 1975; White and Wells 1981). Plugging of surface pores and increased water repellency following fire may diminish water infiltration rate and increase the relative amount of runoff (Krammes and DeBano 1965). Increased overland flow coupled with the loss of soil binding by root systems may result in increased rill and sheet erosion as well as the facilitation of debris flows (Swanson 1981; Wells 1987). This episodic removal of upper soil layers is largely responsible for the undifferentiated character of Mediterranean-climate soils on steep slopes.

The impact of leaching and erosion on overall nutrient budgets depends heavily on the rates and patterns of postfire vegetation recovery. Uptake of nutrients by successional vegetation may significantly diminish nutrient loss. Furthermore, the recovery of leaf area is critical to the reestablishment of prefire hydrologic conditions (Knight et al. 1985).

As indicated above, the relative importance of various processes responsible for nutrient loss (i.e., gasification, volatilization, etc.) varies among nutrients. Gasification may result in considerable losses of nitrogen and sulfur, moderate losses of phosphorus, and only small losses of other mineral elements. Vaporization and convection, on the other hand, account for considerable losses of phosphorus, potassium, calcium, and magnesium (Raison et al. 1985). In general, nitrogen losses are directly proportional to the quantity of fuel consumed by fire, whereas losses of other elements do not increase in direct proportion to fuel consumption or fire intensity (Raison et al. 1985; Binkley and Christensen 1991). Where fire stimulates nitrification, as is often the case in Mediterranean-climate ecosystems, nitrogen losses may be greater than would be predicted based on fuel consumption alone. In general, losses of nutrients to leaching and erosion increase with increased slope and with factors that regulate overland water flow following fire.

Patterns of Nutrient Cycling and Availability

Fire has been observed to cause an increase in mobility and availability of a variety of nutrients, despite overall loss of nutrient capital. These changes arise, in large part, as a consequence of five mechanisms: (1) direct mineral addition in ash; (2) decreased plant uptake; (3) increased microbial activity and decomposition; (4) altered patterns of adsorption and immobilization; (5) changes in oxidation-reduction transformations. In general, the specific mechanism determines the pattern and duration of changes in postfire nutrient mobility and transformations.

Ash nutrient concentrations and nutrient flux in ash fall in a typical chaparral fire are indicated in Table 5–1. Although ash may contain considerable quantities of mineral nutrients such as nitrogen, phosphorus, and potassium, these are not necessarily present in available or exchangeable forms. For example, total nitrogen accounts for nearly 2% of the total ash weight, but exchangeable ammonium and water-soluble nitrate represent considerably less than 1% of that total nitrogen. As expected, the concentration of inorganic mineral nutrients in ash increases with increasing extent of combustion (Raison 1979). Christensen and Muller (1975a) argued that the almost-immediate increase in soil ammonium following fire indicates that this nutrient species increases as a direct consequence of ash addition. The more gradual increase in availability of other nutrient species such as nitrate and phosphate suggested that their increase was a consequence of nutrient transformations that occur subsequent to ash fall. Even though large quantities of ash nutrients are not present in their mineralized form, ash does represent an important pool of relatively decomposable mineral nutrients in soils following fire (Christensen and Muller 1975a; Binkley and Christensen 1991).

Table 5–1. Chemical Composition ($\mu g \cdot g^{-1}$) of Ash Prepared from Combustion of Aboveground Stems and Leaves of *Adenostoma fasciculatum* (chamise). Estimates of Ash Inputs ($kg \cdot ha^{-1}$) are Based on Estimated Ashfall of $3000 \, kg \cdot ha^{-1}$

	Concentration	Ash input to soil
Total carbon	1.7×10^5	509.0
Total nitrogen	7.1×10^3	21.0
Ammonium-N	127.0	0.44
Nitrate-N	1.3	0.004
Total phosphorus	354.0	1.20
Phosphate-P	16.4	0.055
Sulfate	1100	3.7
Total potassium	15.1×10^3	51.0

Diminished competition among plants, especially in the few months immediately following fire, may increase nutrient availability. Christensen and Muller (1975b) and Swanck and Oechel (1991) demonstrated that competition for nutrients in unburned chaparral may be severe and is one of several factors limiting seedling establishment in undisturbed chaparral communities. Christensen (1987) argued that a component of the increased establishment and survival of plants in cleared but unburned chaparral plots (e.g., McPherson and Muller 1969) may be attributed to increased nutrient availability owing to diminished competition. To date, however, trenching experiments specifically designed to evaluate this effect have not been performed.

The impact of fire on decomposition and microbial activity appears to be quite variable. In general, fire has been found to have little effect on decomposition of surface litter and humus, whereas oxidation of organic matter in the upper few centimeters of mineral soil is usually accelerated (Binkley and Christensen, 1991). Christensen and Muller (1975a) documented increased numbers and activity of fungi and bacteria in burned versus unburned chaparral soils. Dunn et al. (1979) demonstrated that the specific character of change in microbial populations following fire (i.e., relative numbers of fungi, heterotrophic bacteria, and nitrifying bacteria) was dependent on fire intensity and soil moisture conditions at the time of burning. Arianoutsou-Faraggitaki and Margaris (1982) found that soil microbial activity in burned phryganic ecosystems of the eastern Mediterranean region was not significantly different from that of unburned areas.

A number of features of the postfire shrubland environment might favor increased microbial activity and decomposition. Increased insolation of soil surfaces may accelerate rates of microbial respiration and decomposition (Christensen and Muller 1975a). In addition, elevated pH caused by ash addition may create conditions that are more favorable to microbial activity. As indicated above, ash addition may increase the availability of reduced carbon substrates from microbes. Finally, Christensen (1973) and Kaminsky (1981) have argued that microbial activity might increase following chaparral fire as a consequence of the removal of the source of antibiotic alleochemicals. This last effect is discussed in more detail later.

The potential role of microbes in the immobilization of particular nutrient species is demonstrated in the seasonal variation in concentrations of soil ammonium. NH_4-N tends to accumulate during the summer months when microbial populations are at their lowest concentrations (Christensen and Muller 1975a; Arianoutsou-Faraggitaki and Margaris 1982). When microbial populations are at a peak in late winter and early spring, ammonium concentrations are lowest. Such seasonal variations in ammonium are absent in soils collected from recently burned shrubland ecosystems (Christensen and Muller 1975a; Arianoutsou-Faraggitaki and

Margaris 1982). However, McKee (1982) and Wilbur and Christensen (1985) have shown that intense fires or increased postfire desiccation, which diminish microbial activity, can increase the availability of phosphorus. The importance of this effect has not, however, been studied in Mediterranean-climate soils.

Rates of nitrification and concentrations of nitrate have generally been observed to increase following fire (Christensen 1973; Christensen and Muller 1975a; Dunn et al. 1979; Arianoutsou-Faraggitaki 1984). These increases have generally been attributed to elevated postfire H, increased availability of ammonium and, perhaps, increased postfire soil aeration (Binkley and Christensen 1991). To the extent that nitrate is a preferred source of soil nitrogen, such changes may increase nitrogen availability to plants. On the other hand, it can be argued that these changes may actually increase overall loss of nitrogen from the ecosystem owing to the increased mobility of nitrate (i.e., increased leaching) and the potential for increased denitrification.

Changes in Nutrient Capital and Nutrient Cycling Patterns in Interfire Years

Marion and Black (1988) studied changes in available nitrogen and phosphorus as well as their distribution among sites on different slope aspects and along a chronosequence extending nearly 90 years following fire. Mineralizable nitrogen generally increased with stand age whereas mineralizable phosphorus generally decreased. The exact trajectory of these changes varied significantly between slope aspects. On average, nitrogen was more available on north-facing slopes than on south-facing slopes, whereas the reverse was true for phosphorus. These researchers attributed these results to differences in the relative importance of nitrogen-fixing shrubs (especially members of the genus *Ceanothus*) on north-facing slopes. The N/P ratio in plant leaf tissues increased with time following fire and was consistently higher on north-facing slopes than on south-facing slopes. These variations undoubtedly affect patterns of decomposition among different age stands and among stands located on different slope aspects; they may also affect the amount of nitrogen and phosphorus volatilized during fires and postfire patterns of nutrient deposition (Marion and Black 1988).

Postfire increases in nutrient availability are often spatially heterogeneous and temporally relatively short lived. Christensen and Muller (1975a) and Stock and Lewis (1986) found that increased nutrient availability was confined primarily to the upper decimeter of the soil profile. Rice (1990) demonstrated that these changes in availability were spatially quite variable on a scale of approximately 10 to 20 m. This spatial variability was correlated with variations in slope and aspect, as well as with patterns of ash deposition on approximately the same spatial scale. Christensen and Muller (1975a) and Stock and Lewis (1986) found that many of the changes in nutrient availability were confined to the first, and sometimes, the second postfire growing season. Differences rapidly diminish in subsequent years and nutrient availability may actually be at its lowest during the first decade following fire than at any other time in the fire cycle (Specht 1969; Marion and Black 1988).

Given the potential nutrient loss associated with burning, there is surprisingly little information available with respect to postfire patterns of nutrient recovery. Figure 5–4 illustrates what is known about inputs and outputs of nitrogen over a complete fire cycle in California chaparral. Variations in precipitation inputs arise as a consequence of the proximity of the maritime and anthropogenic sources of nitrogen-containing aerosols (Schlesinger et al. 1982a). Variations in the amount of symbiotic nitrogen fixation occur largely as a consequence of variations in site conditions. Ellis and Kummerow (1989) and Conard et al. (1985) argue that high densities of nitrogen-fixing shrub seedlings early in the postfire succession accelerate the recovery of nitrogen capital. However, Riggan et al. (1988) point out that shrublands composed of nitrogen-fixing species have considerably higher fuel nitrogen contents and thus are prone to lose more nitrogen to volatilization during fire. It should be noted that claims of

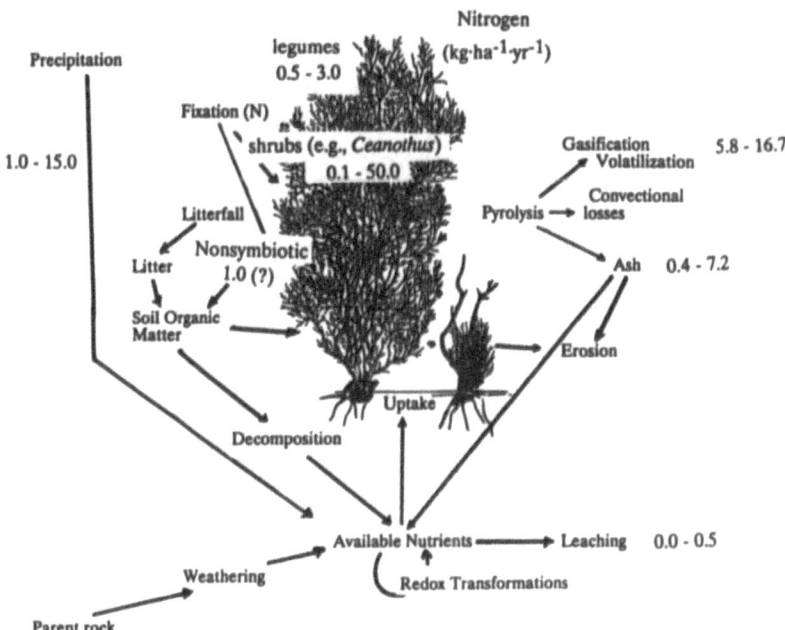

Figure 5–4. Inputs and outputs of N during a fire cycle in California chaparral. (From DeBano and Dunn [1981] and Schlesinger, Gray, and Gilliam [1982].)

greater nonsymbiotic nitrogen fixation immediately following fire have not been substantiated (Binkley and Christensen 1991).

There is a reciprocal relationship between patterns of postfire vegetation recovery and nutrient cycling. The vegetation in immediate postfire years is often characterized by a variety of ephemeral herbs and suffrutescent shrubs in Mediterranean-climate shrublands. The growth and establishment of these plants are clearly affected by availability of nitrogen and phosphorus (Vlamis and Gowan 1961; Christensen and Muller 1975a). However, Rundel and Parsons (1979) found that nutrient uptake by these ephemeral species may diminish the amount of nutrient loss to leaching and thus may be an important mechanism of nutrient conservation. Specht (1969) showed that long-term changes in soil nitrogen and phosphorus were closely tied to patterns of vegetation change in Australian heath.

Fire-Caused Changes in Soil Biochemistry

Soils may contain large quantities of plant secondary chemicals. Some of these chemicals are directly volatilized (Muller 1966), leached (McPherson et al. 1971), or are the breakdown products of the ligneous sclerophyllous litter produced by shrubs in many Mediterranean-climate regions (Kaminsky 1981). Changes in the importance of these chemicals in soils during the fire cycle have been implicated as the mechanisms responsible for changes in soil fertility (Naveh 1960; Muller 1966). In general, it is suggested that toxic allelochemicals accumulate beneath vegetation in interfire years and thereby inhibit germination and growth of potential invading plants. Fire may remove the source of these toxic compounds or directly destroy them in the soil, with the result that postfire soils are more favorable for establishment of these plant species. Keeley and Keeley (1989) raised serious questions regarding the evidence upon which this hypothesis is based. Furthermore, Christensen and Muller (1975a) found that the concentration of many of the putative toxins identified by McPherson et al. (1971) is actually greater in recently burned than in unburned soil. Several studies have demonstrated that a variety of factors, including nutrient availability, innate dormancy of seeds, herbivory, and soil moisture, contribute to the dearth of germination and growth beneath Mediterranean-shrub canopies (Christensen and Muller 1975b; Keeley and Keeley 1989; Swanck and Oechel 1991).

Some secondary chemicals may actually stimulate germination in some species. Wicklow (1977), Jones and Schlesinger (1980), and Keeley et al. (1985) found that chemicals leached from charred wood may stimulate the germination of several chaparral herbs. The specific physiologic mechanism involved in this effect is not yet understood.

Conclusion

The potential impact of fire on the physical and chemical characteristics of soils is magnified by the steep terrain and infertile soils characteristic of many Mediterranean-climate ecosystems. Fires ordinarily are not sufficiently hot to alter soil properties such as texture or minerology; however, loss of organic matter and development or migration of hydrophobic layers may greatly alter soil hydrologic properties. These changes, coupled with loss of plant cover and exposure of soil to direct impact by raindrops, may greatly alter ecosystem hydrologic budgets and increase soil erosion. Loss of mineral nutrients from soils and fuels occurs as a result of oxidation of compounds to gaseous form, vaporization, ash convection, mobilization and leaching, and acclerated erosion. The relative importance of these various processes depends on nutrient species in question, characteristics of the soil, and variations in fire behavior.

Despite such losses, available forms of many nutrients typically increase following fire, contributing to often-observed postfire flushes of plant growth. Increased availability of nutrients following fire results from direct nutrient inputs in ash, decreased plant uptake, changes in patterns of nutrient adsorption and immobilization, increased mineralization, and altered patterns of nutrient transformation. Although N and P are usually present in high concentration in ash, their ultimate availability depends on subsequent mineralization. Except where fires have been especially intense, microbial activity and decomposition are generally higher in burned areas, resulting in higher rates of mineralization and transformations such as nitrification.

Recovery of mineral capital lost to burning occurs primarily via atmospheric inputs for most nutrients. However, increased symbiotic nitrogen fixation following burning is important in many regions. Evidence for accelerated rates of nonsymbiotic nitrogen fixation is lacking. Although postfire plant recovery may be accelerated by increased nutrients, such regrowth may also minimize losses of nutrients from Mediterranean ecosystems.

Specific patterns of nutrient change vary considerably among and within fires. Such spatial variation may play a critical role in the maintenance of ecosystem diversity.

References

Arianoutsou-Faraggitaki, M. 1984. Post-fire successional recovery of a phyrganic (east Mediterranean) ecosystem. *Acta Oecologica Plantarum* 5:387–394.

Arianoutsou-Faraggitaki, M., Margaris, N.S. 1981. Fire-induced nutrient losses in a phryganic (east Mediterranean) ecosystem. *International Journal of Biometeorology* 25:341–347.

Arianoutsou-Faraggitaki, M., Margaris, N.S. 1982. Decomposers and the fire cycle in a phryganic (east Mediterranean) ecosystem. *Microbial Ecology* 8:91–98.

Beatty, S.W. 1989. Fire effects on soil heterogeneity beneath chamise and redshanks chaparral. *Physical Geography* 10:44–52.

Binkley, D., Christensen, N.L. 1991. The effects of canopy fire on nutrient cycles and plant productivity. *In* S.R. Lavin and P. Ommi (eds.), *High Intensity Fire*. Academic Press (in press).

Boerner, R. 1982. Fire and nutrient cycling in temperate ecosystems. *BioScience* 32:187–192.

Boring, L.R., Swank, W.T., Waide, J.B., Henderson, G.S. 1988. Sources, fates, and impacts of nitrogen inputs to terrestrial ecosystems: Review and synthesis. *Biogeochemistry* 6:119–159.

Boucher, C., Moll, E.J. 1981. South African Mediterranean shrublands, pp. 233–247. *In* F. di Castri, D.W. Goodall, and R.L. Specht (eds.), *Ecosystems of the World: Vol. II, Mediterranean-Type Shrublands*. Elsevier Scientific Publishing, New York.

Christensen, N.L. 1973. Fire and the nitrogen cycle in California chaparral. *Science* 181:66–68.

Christensen, N.L. 1985. Shrubland fire regimes and their evolutionary consequences, pp. 85–100. *In* S.T.A. Pickett and P.S. White (eds.), *The Ecology of Natural Disturbance and Patch Dynamics*. Academic Press, New York.

Christensen, N.L. 1987. The biogeochemical consequences of fire and their effects on the vegetation of the coastal plain of the southeastern United States, pp. 1–21. *In* L. Trabaud (ed.), *The Role of Fire in Ecological Systems*. SPB Academic Publishing, The Hague, The Netherlands.

Christensen, N.L., Muller, C.H. 1975a. Effects of fire on factors controlling plant growth in *Adenostoma* chaparral. *Ecological Monographs* 45:29–55.

Christensen, N.L., Muller, C.H. 1975b. The relative importance of factors controlling germination and seedling survival in *Adenostoma* chaparral. *American Midland Naturalist* 93:71–78.

Conard, S., Jaramill, A., Cromack, K., Rose, S. 1985. The role of the genus *Ceanothus* in western forest ecosystems. USDA Forest Service Gen. Tech. Report PNW–182.

Connaughton, C.A. 1935. Forest fire and accelerated erosion. *Journal of Forestry* 33:751–752.

Cotton, F.A., Wilkinson, G. 1988. *Advanced Inorganic Chemistry*, 5th ed. Wiley, New York.

DeBano, L.F. 1982. Assessing the efects of management actions on soild and mineral cycling in Mediterranean ecosystems. *In* C.E. Conrad and W.C. Oechel (tech. coord.), *Dynamics and Management of Mediterranean-Type Ecosystems*. USDA Gen. Tech. Report PSW–58, pp. 345–350.

DeBano, L.F., Conrad, C. 1978. The effect of fire on nutrients in a chaparral ecosystem. *Ecology* 59:489–497.

DeBano, L.F., Eberlein, G.E., Dunn, P.H. 1979. Effects of burning on chaparral soils: I. Soil nitrogen. *Soil Science Society of America Journal* 43:504–509.

di Castri, F. 1981. Mediterranean-type shrublands of the world, pp. 1–52. *In* F. di Castri, D.W. Goodall, and R.L. Specht (eds.), *Ecosystems of the World. Vol. II, Mediterranean-Type Shrublands*. Elsevier Scientific Publishing, New York.

Dunn, P.H., DeBano, L.F., Eberlein, G.E. 1979. Effects of burning on chaparral soils: II. Soil microbes and nitrogen mineralization. *Soil Science Society of America Journal* 43:509–514.

Ellis, B.A., Kummerow, J.K. 1989. The importance of N_2 fixation in *Ceanothus* seedlings in early postfire chaparral, pp. 115–116. *In* S.C. Keeley (ed.), *The California Chaparral: Paradigms Reexamined*. Natural History Museum of Los Angeles County, Los Angeles, CA.

Greenwood, N., Earnshaw, A. 1984. *Chemistry of the Elements*. Pergamon, Oxford, England.

Grier, C. 1975. Wildfire effects on nutrient distribution and leaching in a coniferous ecosystem. *Canadian Journal of Forest Research* 5:599–607.

Harwood, C., Jackson, W. 1975. Atmospheric losses of four plant nutrients during a forest fire. *Australian Forestry* 38:92–99.

Jones, C.S., Schlesinger, W.H. 1980. *Emmenanthe penduliflora* (Hydrophyllaceae): Further considerations of germination response. *Madroño* 27:122–125.

Kaminsky, R. 1981. The microbial origin of the allelopathic potential of *Adenostoma fasciculatum* H. and A. *Ecological Monographs* 51:365–382.

Keeley, J.E., Keeley, S.C. 1989. Allelopathy and the fire-induced herb cycle, pp. 65–72. *In* S.C. Keeley (ed.), *The California Chaparral: Paradigms Reexamined*. Natural History Museum of Los Angeles County, Los Angeles, CA.

Keeley, J.E., Morton, B.A., Pedrosa, A., Trotter, P. 1985. The role of allelopathy, heat, and charred wood in the germination of chaparral herbs and suffrutescents. *Journal of Ecology* 73:445–458.

Knight, D.H., Fahey, T.J., Running, S.W. 1985. Water and nutrient outflow from contrasting lodgepole pine forests in Wyoming. *Ecological Monographs* 55:29–48.

Krammes, J.S., DeBano, L.F. 1965. Soil wettability: A neglected factor in watershed management. *Water Resources Research* 1:283–286.

Kutiel, P., Naveh, Z. 1987. Soil properties beneath *Pinus halepansis* and *Quercus calliprinos* trees of burned and unburned mixed forest on Mt. Carmel, Israel. *Forest Ecology and Management* 20:11–24.

Longstreth, D.J., Patten, D.T. 1975. Conversion of chaparral to grass in central Arizona: Effects on selected ions in watershed runoff. *The American Midland Naturalist* 93:25–34.

McKee, W.H. Jr. 1982. Changes in soil fertility following prescribed burning on coastal plain pine sites. USDA Forest Service Research Paper SE–234.

McPherson, J.K., Muller, C.H. 1969. The allelopathic effects of *Adenostoma fasciculatum*, "chamise," in the California chaparral. *Ecological Monographs* 39:177–198.

McPherson, J.K., Chow, C.H., Muller, C.H. 1971. Allelopathic constituents of the chaparral shrub *Adenostoma fasciculatum*. *Phytochemistry* 10:2925–2933.

Malanson, G.P. 1984. Fire history and patterns of Venturan subassociations of California coastal scrub. *Vegetatio* 57:121–123.

Margaris, N.S. 1976. Structure and dynamics in a phryganic (east Mediterranean) ecosystem. *Journal of Biogeography* 3:249–259.

Margaris, N.S. 1981. Adaptive strategies in plants dominating mediterranean-type ecosystems. Pages 309–316. *In* F. di Caseri, D.W. Goodall, and R.L. Specht (eds.), Ecosystems of the World. Vol. 11: Mediterranean-Type Shrublands. Elxevier, Amsterdam, p. 643.

Margaris, N.S., Adamandiadou. S., Siafaca, L., Diamantopoulos, J. 1984. Nitrogen and phosphorus content in plant species of Mediterranean ecosystems in Greece. *Vegetatio* 55:29–35.

Marion, G.M. 1982. Nutrient mineralization processes in Mediterranean-type ecosystems. *In* C.E. Conrad and W.C. Oechel, *Dynamics and Management of Mediterranean-Type Ecosystems*. USDA Gen. Tech. Report PSW–58, pp. 313–320.

Marion, G.M., Black, C. 1988. Potentially available nitrogen and phosphorus along a chaparral fire cycle chronosequence. *Soil Science Society of America Journal* 52:1155–1162.

Megahan, W.F., Molitor, D.C. 1975. Erosional effects of wildfire and logging in Idaho. *Watershed Management Symposium*. American Society of Civil Engineering.

Minnich, R.A. 1983. Fire mosaics in southern California and northern Baja California. *Science* 219:1287–1294.

Minnich, R.A. 1988. Chaparral fire history in San Diego County and adjacent northern Baja California: An evaluation of natural fire regimes and the effects of suppression management, pp. 37–48. *In* S.C. Keeley (ed.), *The California Chaparral: Paradigms Reexamined*. Natural History Museum of Los Angeles County, Los Angeles, CA.

Muller, C.H. 1966. The role of chemical inhibition (allelopathy) in vegetation composition. *Bulletin of the Torrey Botanical Club* 93:332–351.

Muller, C.H., Hanawalt, R.B., McPherson, J.K. 1968. Allelopathic control of herb growth in the fire cycle of California chaparral. *Bulletin of the Torrey Botanical Club* 95:225–231.

Naveh, Z. 1960. The ecology of chamise as affected by its toxic leachates. *Bulletin of the Ecological Society of America* 41:56–57 (Abstract).

Olson, J.S. 1981. Carbon balance in relation to fire regimes, pp. 327–378. *In* H.A. Mooney, T.M. Bonnicksen, N.L. Christensen, J.E. Lotan, and W.A. Reiners (eds.), *Fire Regimes and Ecosystem Properties*. USDA Forest Service Gen. Tech. Report WO–26.

Quézel, P. 1981. Floristic composition and phytosociological structure of sclerophyllous matorral around the Mediterranean, pp. 107–122. *In* F. di Castri, D.W. Goodall, and R.L. Specht (eds.), *Ecosystems of the World: Vol. II, Mediterranean-Type Shrublands*. Elsevier Scientific Publishing, New York.

Raison, R.J. 1979. Modification of the soil environment by vegetation fires, with particular reference to nitrogen transformations: A review. *Plant and Soil* 51:73–108.

Raison, R.J., Khanna, P., Woods, P. 1985. Mechanisms of element transfer to the atmosphere during vegetation fires. *Canadian Journal of Forest Research* 15:132–140.

Rice, S. 1990. Impact of fire on spatial variation in soil and vegetation in Sierra Nevada chaparral. Unpublished M.S. Thesis, Duke University, Durham, NC.

Riggan, P.J., Goode, S., Jacks, P.M., Lockwood, R.N. 1988. Interaction of fire and community development in chaparral in southern California. *Ecological Monographs* 58:155–176.

Rundel, P.W. 1979. Ecological impact of fires on mineral and sediment pools and fluxes, pp. 17–21. *In* J.K. Agee (ed.), *Fire and Fuel Management in Mediterranean Climate Ecosystems*. Man and the Biosphere Technical Note No. 2.

Rundel, P.W. 1981. Structural and chemical components of flammability, pp. 183–207. *In* H.A. Mooney, T.M. Bonnicksen, N.L. Christensen, J.E. Lotan, and W.A. Reiners (eds.), *Fire Regimes and Ecosystem Properties*. USDA Forest Service Gen. Tech. Report WO–26.

Rundel, P.W., Parsons, D.J. 1979. Structural changes in chamise (*Adenostoma fasciculatum*) along a fire-induced age gradient. *Journal of Range Management* 32:462–466.

Schlesinger, W.H., Gray, J.T., Gill, D.S., Mahall, B.E. 1982a. *Ceanothus megacarpus* chaparral: A synthesis of ecosystem processes during development of annual growth. *The Botanical Review* 48:71–117.

Schlesinger, W.H., Gray, J.T., Gilliam, F.S. 1982. Atmospheric deposition processes and their importance as sources of nutrients in a chaparral ecosystem of southern California. *Water Resources Research* 18:623–629.

Specht, R.L. 1969. A comparison of the sclerophyllous vegetation characteristic of Mediterranean-type climates in France, California, and southern Australia:

II. Dry matter, energy, and nutrient accumulation. *Australian Journal of Botany* 17:293–308.

Specht, R.L. 1981. Mediterranean ecosystems in southern Australia, pp. 203–232. *In* F. di Castri, D.W. Goodall, and R.L. Specht (eds.), *Ecosystems of the World: Vol. II, Mediterranean-Type Shrublands*. Elsevier Scientific Publishing, New York.

Stark, N.M. 1977. Fire and nutrient cycling in a Douglas fir/larch forest. *Ecology* 58:16–30.

Stock, W.D., Lewis, O.A.M. 1986. Soil nitrogen and the role of fire as a mineralizing agent in a South African coastal fynbos ecosystem. *Journal of Ecology* 74:317–328.

Swanck, S.E., Oechel, W.C. 1991. Interactions among the effects of herbivory, competition, and resource limitation on chaparral herbs. *Ecology* 72:104–115.

Swanson, F.J. 1981. Fire and geomorphic processes, pp. 421–444. *In* H.A. Mooney, T.M. Bonnicksen, N.L. Christensen, J.E. Lotan, and W.A. Reiners (eds.), *Fire Regimes and Ecosystem Properties*. USDA Forest Service Gen. Tech. Report WO–26.

Tiedemann, A.R., Helvey, J.D., Anderson, T.D. 1978. Stream chemistry and watershed nutrient economy following wildfire and fertilization in eastern Washington. *Journal of Environmental Quality* 7:580–588.

Turner, M.G. 1987. Landscape heterogeneity and disturbance. *Ecological Studies*, Vol. 64. Springer-Verlag, New York.

Turner, M.G., Romme, W.H. 1994. Landscape dynamics in crown fire ecosystems. *Landscape Ecology*.

Vlamis, J., Gowan, K.D. 1961. Availability of nitrogen, phosporus, and sulfur after brush burning. *Journal of Range Management* 14:38–40.

Weast, R. 1982. *CRC Handbook of Chemistry and Physics*. CRC Press, Boca Raton, FL.

Wells, S.G. 1987. The effects of fire on the generation of debris flows in southern California. *Geological Society of America, Reviews in Engineering Geology* 7:105–114.

White, W.D., Wells, S.G. 1981. Geomorphic effects of the La Mesa Fire, pp. 73–90. *The La Mesa Fire Symposium*. Los Alamos National Laboratory LA-9236-NERP.

Wilbur, R.B., Christensen, N.L. 1985. Effects of fire on nutrient availability in a North Carolina coastal plain pocosin. *The American Midland Naturalist* 110:54–61.

6. Fire and Water Yield: A Survey and Predictions for Global Change

Serge Rambal

By the late 1800s it was apparent to some that burning (or clearing) the vegetation increased water flow and accelerated flood damage (Hibbert et al. 1974). However, these phenomena had already been described at the beginning of our era by Pliny the Elder in his *Naturalis Historia* (in Lieuthagi 1972). The first experiments to determine the role of vegetation on water yield began in 1900 in the Emmenthal Mountains, Switzerland, and in 1909 in Wagon Wheel, CO, USA. The effect of wildfire on streamflow was first documented by Hoyt and Troxell in 1932. They found that burning chaparral caused the average annual streamflow of Fish Creek, CA, to increase 29%, or about 40 mm. In addition they found that peak discharges and sediment loads carried by the streams also increased. The same results have since been obtained by many other authors (e.g., Anderson 1949).

In a context in which fire can be used for vegetation conversion, it is necessary for water and land managers to weigh the benefits of the extra water against the cost of damage and determine the best alternative for vegetation management in water-short areas.

In this chapter, I discuss only the role of fire on the annual water yield and, consequently, on the total evaporation from soil and plant canopy. I will not consider here the adverse effects on flood discharges or on soil loss, processes occuring at shorter time scales and concerning smaller amounts of water.

Fire and Annual Water Yield

The basic relationship used in quantifying the various aspects of water balance on a watershed scale is the hydrological equation: $P = Q + E$ where P is precipitation, Q is the streamflow, also called water yield in this chapter and E is evaporation. P, Q, and E are generally expressed in millimeters of water. The volumetric yield may be obtained by multiplying Q by the area of the watershed. Actual evaporation (E) includes evaporative losses via transpiration, interception, and evaporation from the soil surface. In view of the difficulties associated with the direct measurement of E on a watershed scale, E is generally determined as a residual in the above equation. An implicit assumption is that no liquid water enters the watershed except as precipitation or lost except through streamflow. The water balance equation has the advantage of providing data from large areas and integrating the variability occuring across landscapes. However, this advantage is offset by the disavantage that its degree of accuracy is often low and that it is difficult to obtain valid estimates of E for periods of less than a year. The following results concern only annual water balances and try to establish empirical relationships between vegetation changes and streamflow.

There is a widespread and ancient belief, perhaps since Plato (2400 B.P.), that the destruction of vegetation has caused wells, springs, and even rivers to cease flowing, at least during the dry seasons. In this context, the paper of Trimble et al. (1963) entitled "Cutting the forest to increase water yield" could appear slightly iconoclastic. As Hamilton (1988) said, "Policies establishing protection forests are being advocated because of a supposed 'sponge' effect of the tree roots. It is claimed that the roots soak up water in the wet periods and release it slowly and evenly in the dry season to keep water supplies adequately restored. Roots may be more appropriately labelled a 'pump' rather than a 'sponge'. They certainly do not release water in the dry season, but rather remove it from the soil in order that the trees may transpire." It is difficult to reconcile such policies that protect forest and watershed experiments that indicate most universally an increased annual water yield following logging or burning forests.

In 1965, Shachori and Michaeli reviewed 157 controlled watershed or small-plot studies from many parts of the world. Their analysis indicated smaller water yields from forest, woodland, and chaparral than from grasslands or bare areas. A significant linear relationship exists between annual streamflow and annual rainfall for two cover types (see Figure 6–1). For forest, woodland, and chaparral: $Q = 0.805 (P - 398)$ and for grasslands and bare areas: $Q = 0.920 (P - 281)$ where Q and P are the annual water yield and precipitation in millimeters, respectively.

All burning experiments quoted in the Shachori and Michaeli paper have shown increases in water yield prior to vegetation recovery. The

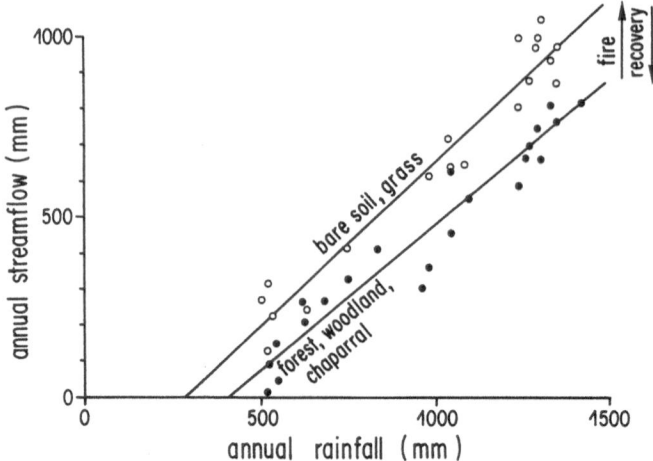

Figure 6–1. Relationships of the water yield (or annual streamflow) of two cover types with varying precipitation. (Redrawn from Shachori and Michaeli 1965.)

average difference between the two regression lines, in the 800 to 1000 mm annual rainfall range, is approximately 120 mm. This last value can be considered as a predictive quantification of the amount of water yield increase.

Hibbert (1967) summarized the results of 39 published experiments (34 in the United States, 5 elsewhere), some of which had been already reviewed in Shachori and Michaeli. All showed an increase in water yield following forest reduction, but the annual increases ranged from a few millimeters to approximately 450 mm. Consequently, a practical upper limit of water yield increase appears to be 4.5 mm per year for each percent reduction in forest cover. Hibbert (1967) made the following generalizations: "(1) reduction of forest cover increases water yield, (2) establishment of forest cover on sparsely vegetated land decreases water yield and (3) response to treatment is highly variable and for the most, unpredictable."

This review has been updated by Bosch and Hewlett (1982) using the results from 55 watershed experiments from around the world. Empirical analysis shows that increases in water yield can be ranked according to the percentage of reduction in vegetative cover. Coniferous forest gave the largest increase in water yield with decreasing cover, followed by deciduous hardwood or mixed hardwood and finally, by scrub vegetation (Figure 6–2). The change in annual water yield per percent change in vegetation cover dQ/dC, are 4, 2.5, and 1 mm respectively for these three cover types. Increases in water yield were greater in higher rainfall areas. However, predicting changes in water yield using these empirical results is problematic. The variances in their data are large and, as McNaughton and Jarvis (1983) pointed out, rainfall and vegetation types are related. Thus in all the experiments scrub vegetation was associated with low rainfall, and coniferous forest was generally associated with higher rainfall.

A well known contradiction to those patterns suggested above is worth mentioning. Melbourne, a city of 3 million people on the southern coast of Australia, is supplied with water from forested watersheds in the mountains to the north and east of the city. These watersheds occupy an area of about $1000 \, km^2$ and supply an average of $375 \, Mm^3 a^{-1}$. The eucalypt forests covering these watersheds can be classified as either ash-type or mixed species. The ash-type forests (dominated by *Eucalyptus regnans*), are of special significance because the area occupied by these forests, though only 50% of the total watershed area, produces approximately 80% of total streamflow. In January 1939, a disastrous wildfire swept through the Melbourne's watersheds. In the decade following the 1939 fire, evidence began to accumulate that there was a relationship between the age of an *Eucalyptus regnans* forest and the streamflow it yielded (Langford 1976). Langford found that the replacement of a largely mature ash-type forest with vigorous regrowth forest induced a long-term reduction in streamflow. For instance, streamflow was reduced by 24% where some 50% of the area was converted into regrowth forest following the fire.

Contrary to all the examples reported earlier, the long-term streamflow trends showed a decline starting 3 to 5 years after the fire and reached a maximum 15 to 20 years after the fire. Why did the mature ash-type forest consume less water than the regrowth forest? The mechanisms of

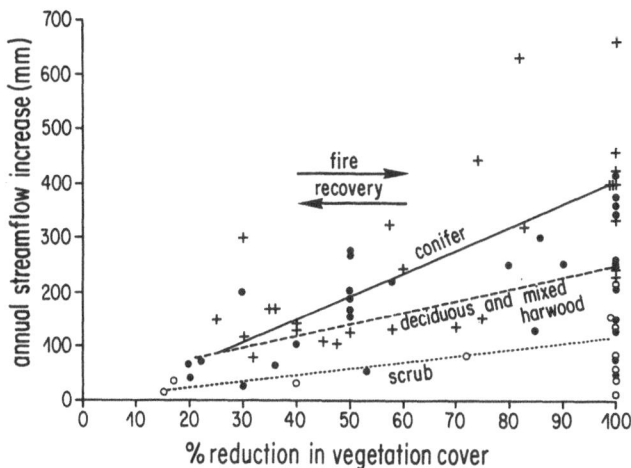

Figure 6–2. Increase of the water yield associated with the reduction of the vegetation cover. Key: scrubs (empty circles), deciduous, and mixed hardwoods (filled circles), and conifers (crosses). (Redrawn from Bosch and Hewlett 1982.)

reduced water yield proposed by Langford (1976) are (1) fog drip and canopy interception, (2) regulation of transpiration through the internal resistance to liquid water flow (the mature trees have a taller canopy and therefore a lower evaporation because water must be lifted to a greater height) and (3) differences in the root systems. For Holmes and Wronski (1982), energy transfer and aerodynamic factors would be a more likely cause. They showed that for a very tall canopy (*Eucalyptus regnans* is one of the word's tallest hardwoods with the tallest standing tree at approximately 98 m) a significant reduction in transpiration with height could offset the predicted increase in interception loss with height in the long-term water balance.

Estimating Water Yield at the Landscape Level

The wide variation of the observed increases in streamflow demonstrate the difficulty of precisely predicting landscape response to fire. The amount and seasonal distribution of rain interact with soil and site specific conditions to produce the observed responses.

A Chaparral Case

For the chaparral in Arizona, Hibbert et al. (1974) developed a relation-ship, based on experimental results, that is useful for planning purposes because it gives the mean or "normal" expected response over a period of several years. The equation is: $dQ = 0.740 (P - 430)$ where dQ is mean annual increase in streamflow and P is mean annual precipitation. Assuming 560 mm of precipitation, which is the mean rainfall for chaparral in Arizona, the estimated dQ is 95 mm. Hibbert et al. (1974) observed that discounting is necessary when extrapolating these experimental results to larger areas where fire may not be as complete or as continuous. In an economic analysis of the chaparral conversion of the Salt-Verde basin (Arizona), Brown et al. (1974) assumed only 60% to be burnt. The previous equation becomes: $dQ = 0.590 (P - 255)$.

I now present three examples in the case of heterogeneous landscapes. The first considers the role of vegetation type and burning frequency.

The Bosch et al. Model: A Fynbos Case

Bosch et al. (1986) proposed a model for comparing water yield from fynbos watersheds burnt at different time intervals. Fynbos, a sclerophyl-lous vegetation type, occurs primarily in the western and southern Cape, South Africa. Results from experimental work suggest that a fire in fynbos communities, as in most other vegetation types, will cause an immediate increase in water yield followed by a decrease in yield as vegetation recolonize the burned area. The postfire increase in water yield will last for a certain period depending on vegetation type, recovery rate, and climate. Each of these factors should be quantified when considering the effects of different burning cycles.

Changes in water yield following burning cycles are presented schematically in Figure 6–3. Q_0 represents the stabilized flow from a watershed with mature fynbos vegetation of a given type and Q_1 the mean annual streamflow from the same watershed immediately following fire. The line BC represents the decline in water yield as vegetation recovers. If there is no subsequent fire, water yield will decline to the prefire level, after which it will remain steady. The trapezoid A'B'C'D' represents the streamflow response or the cumulative water gain if the watershed was burned after T years (see fire frequency on Figure 6–3). Variations in position, size, and form of A'B'C'D' represent variations in response to the different vegetation types at different fire frequencies. The Bosch et al. (1986) model assumes a linear relationship between streamflow and time since the last fire. This is a simplified representation of the recovery process likely to be sigmoidal. The model also assumes, as in the previous model proposed for the chaparral, a constant mean annual rainfall.

The increase above Q_0 is $dQ = Q_1 - Q_0$, which was experimentally determined to be approximately $180\,\text{mm}\,\text{a}^{-1}$ in the case of a tall closed shrubland in a high rainfall watershed. Bosch et al. (1986) hypothesized that dQ is primarily a function of the standing biomass, which is related to other parameters such as total vegetation cover or leaf area index.

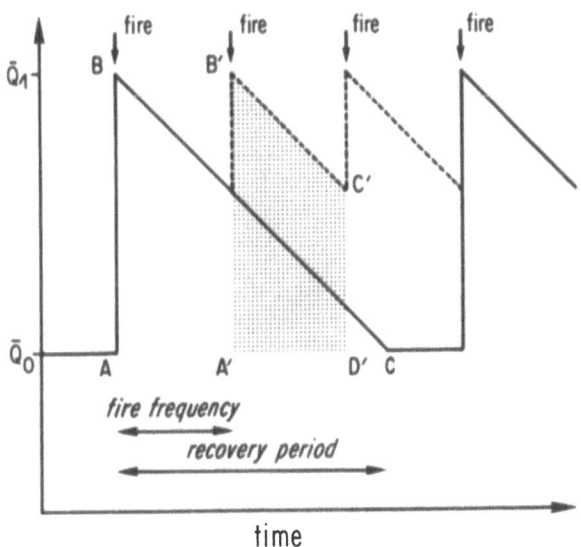

Figure 6–3. Schematic representation of shifts in mean water yield following two burning cycles in fynbos. (Modified from Bosch et al. 1986.)

Structural classes of fynbos were assigned an arbitrary biomass rank from 1 to 7 and a related expected postfire maximum increase in water yield (Figure 6–4A). For instance, tall closed shrublands had a biomass of approximately $5000\,g\,m^{-2}$ and a biomass rank of 1, tall open shrublands had a biomass of $3000\,g\,m^{-2}$ and a rank of 3, and tall mid-dense herblands had a biomass of $500\,g\,m^{-2}$ and a rank of 4. In a *Quercus coccifera* garrigue growing under 980 mm rainfall, I observed an increase in water yield (dQ) of 80 mm (Rambal, unpublished data). Its standing biomass was about $1800\,g\,m^{-2}$. The hypothetical postfire maximum increase for vegetation with a biomass rank between 3 and 4 is 60 to 90 mm.

The rate of canopy recovery will be a major factor in determining reduction in streamflow. The slope of the lines BC or B'C' in Figure 6–3 depicts this reduction. Several factors related to the vegetation determine the slope of this line. Bosch et al. (1986) observed that communities with a large proportion of obligate seeding species recover relatively slowly and, therefore, the slope of BC (or B'C') is slight. A community with a

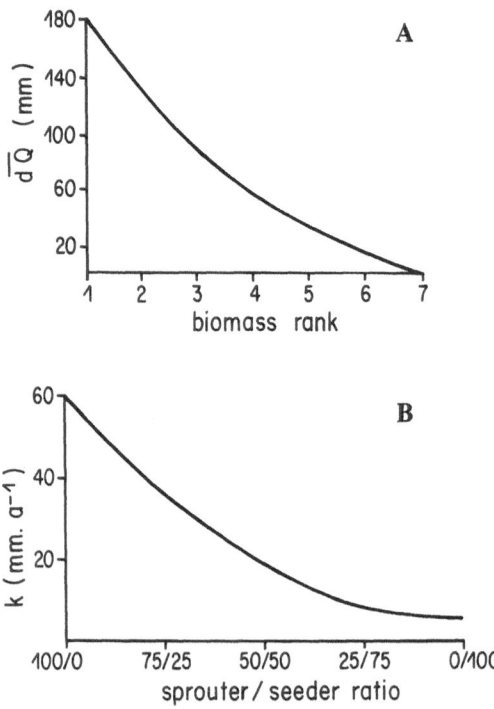

Figure 6–4. (A), Relationship of the shift in mean annual streamflow after fire with the biomass rank of the prefire vegetation cover (1 = high biomass, forest; 7 = low biomass, sparse vegetation); (B), relationship of the annual rate of reduction in the maximum water yield shift and the sprouter/seeder ratio of the vegetation cover.

large proportion of sprouters will cause a steeper decline. A relationship between the expected annual rate of reduction in water yield (k) and the sprouter/seeder ratio is presented in Figure 6–4B. We can also calculate the recovery period by dividing dQ by k. For a tall closed shrubland (rank 1) with mainly sprouting species, the recovery period will be 180/60 or 3 yr and 180/4 or 45 yr in the case of a seeder-dominated community. Applying these values to the already-mentioned *Q. coccifera* dominated garrigue with a sprouter/seeder ratio of approximately 90/10 leads to a recovery period of 80/50 or 1.6 yr. The observed recovery period for the leaf area index is approximately 2 years. For a given community, this ratio will differ depending upon fire frequency (see, for instance, Trabaud 1987). These possible changes should be kept in mind when selecting k for modeling purposes.

Melbourne's Water Supply: The Case of an Ash-Type Forest.

Langford (1976) using multiple regression analysis estimated a regional relationship between forest descriptors and long-term water yield response: $dQ = -(153 + 1.79A - 2.29M)$ where dQ is the average long-term reduction in streamflow or streamflow shift (in mm), A is the percentage of mature forest regenerated by the fire, and M is the percentage of mixed forest growing on the watershed. Ash-type forest regeneration accounted for 75% of the variation in water yield while mixed-forest regeneration accounted for the residual 25%. An increase in ash-type forest regeneration reduced flow while in more mixed-species forest flow reduction was less pronounced. Only four watersheds were used in this calculation, thus the results of the regression analysis should be interpreted with care. For example Ronan and Ducan (1980), cited in Kuczera (1987), observe that Langford's equation predicts a net long-term increase in streamflow of 76 mm for a watershed covered completely by mixed species both before and after a fire!

Kuczera (1987) recently proposed a two-parameter equation to describe the decrease in the water yield: $dQ = -KL_{max}(t - 2)\exp(1 - K(t - 2))$ for $t > 2$ and, $dQ = 0$ for $t < 2$ where t is the time after the fire in years, K is an empirical parameter in $year^{-1}$, 1/K is the time in years from the beginning of the yield reduction to its maximum, and L_{max} is the maximum yield reduction following the fire in mm. From a regional analysis, Kuczera found that L_{max} may be estimated by $L_{max} = 6.15A$ (A has the same meaning as in Langford's equation). 1/K ranges generally from 20 to 30 yr but may attain 60 yr. During the period immediately following fire, Kuczera's equation in its original form does not predict the observed positive yield shift of the streamflow. The extrapolation of this equation for $t < 2$ yr results in a shift which is too large.

Daniell and Kulik (1987) developed a simplified model to estimate the shift in water yield due to fires. Their model is based on the concept of a leaking watershed. It roughly describes the evaporation between rainfall events as a function of root expansion. The soil is divided into two hypothetical soil layers: (1) a streamflow generating layer and 2) a deeper layer generating a streamflow unrecorded by the hydrograph at the streamgage station. During the first few years after the fire (step 1) evaporation is limited by the shallow root system of the young trees. There is, however, more streamflow than before the fire. Subsequent to this period the roots grow enough to extract a sufficient quantity of water for evapotranspiration and all water is taken from the streamflow generating horizon of the soil (step 2). Hence, the loss in postfire is greater than in the prefire conditions. In the last step (step 3), the roots penetrate to the lower soil layers and start to extract water from the unrecorded streamflow. As more water is extracted from the deeper layer and less from the upper one, streamflow increases and water yield returns to prefire level. The model's parameters are the initial root depth immediately following fire, maximal root depth, depth of the streamflow generating soil layer, time to forest maturity, annual evaporation of the mature forest, and an index of the availability of soil water. Figure 6–5 shows the direct comparison of Kuczera's equation and that of Daniell and Kulik's model. I also report a shift in the water yield based on the Bosch et al. (1986) model.

Kuczera (1987) used his formula to show the potential vulnerability of Melbourne's water supply to a major fire. This vulnerability can be

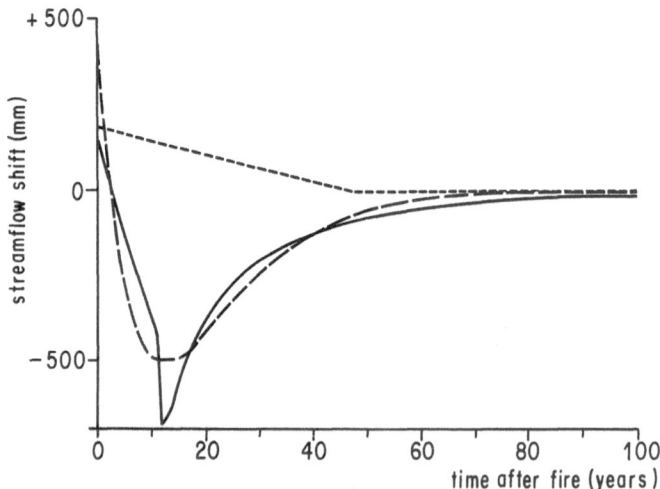

Figure 6–5. Mean annual streamflow following a fire described with: (1) Daniell and Kulik's model (solid line); (2) Kuczera's equation (long dashed line). (From Daniell and Kulik 1987.) The model of Bosch et al. is also reported for comparison (short dashed line).

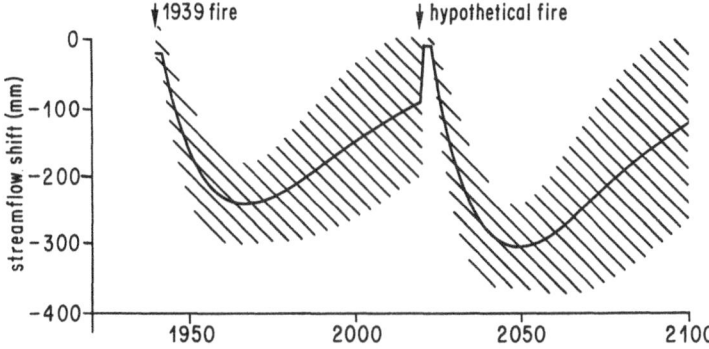

Figure 6–6. Evolution of the mean annual streamflow relative to the prefire yield and its 90% confidence level in the worst case fire scenario for Melbourne's water supply. (Redrawn from Kuczera 1987.)

illustrated by considering the results of a worst case scenario in which two assumptions are made: (1) the next major fire after 1939 will occur around 2020, at which time the average yield of 1939 regrowth ash forest is estimated to be 90 mm below the oldgrowth yield; and (2) the fire will convert all the mature ash-type forest to regrowth.

The expected yield reduction relative to an old-growth forest with its 90% confidence limits is displayed in Figure 6–6. Following the 2020 fire, the expected increase in yield will be about 94 mm. About 25 years after the fire, the expected reduction in average streamflow relative to the 2020 yield is about 210 mm. This represents a loss of water equal to 65% of the Melbourne's 1982 annual consumption. Recovery of the prefire yields takes another 80 yr or more.

A Case With Woodlands and Shrublands in Southern France

Changes in vegetation cover linked to human activities, are particularly obvious in the Mediterranean area in southern France. At the present time, the cover of woody plant communities (garrigues, oak woodlands, *Quercus ilex* coppices) is increasing at the cost of rangelands and crops. In this context, yearly fires burned approximately 43,000 ha for the period 1977 to 1986 (Le Houérou 1987) or 1.3% of the 3340 10^3 ha total land area of "natural vegetation" (shrublands or woodlands).

Vegetation maps and a hydrological simulation model were used in order to estimate the consequences in landscape changes on the water yield of a 35 km^2 karstic watershed near Montpellier, France. Five vegetation maps were established at a scale of 1/50,000. Vegetation physiognomy had been assessed quantitatively through a codification based upon the respective projected covers of trees higher than 2 m, shrubs, and herbaceous layer. Projected cover of the three layers was estimated visually and the boundaries of vegetation types were delineated from five panchromatic aerial photographs obtained in 1946, 1954, 1961, 1971, and 1979. For the purpose of this study, four simplified classes, or vegetation formations, based only on the tree cover were used: shrublands (<25%), open (25 to 50%), mid-dense (50 to 75%) and dense (>75%) woodlands. The cultivated areas (6% in 1946, 0% in 1979) were neglected. Nevertheless, leaf area index is probably the most useful single structural variable for quantifying the energy and water exchanges of different vegetation types. Field measurements at and near the watershed resulted in leaf area index (LAI) ranging from 1.2 to 4.0. We assigned to each vegetation type an averaged LAI value. These corresponded to <1.8, 1.8–2.4, 2.4–3.2, and >3.2 for shrublands, open, mid-dense, and dense woodlands, respectively. In cases of lithologic homogeneity, the vegetation formation and its associated LAI can be used as "hydrologic unit" (Rambal et al. 1985).

In our hydrological simulation model, we took into account one soil layer comprising the entire root system. Actual evaporation varies as a function of (1) potential evaporation; (2) leaf area index; and (3) canopy conductance. Canopy conductance depends solely upon canopy water potential. To satisfy the evaporative demand, the roots extract water from the soil layer. We assume that the canopy generates a potential value just sufficient to equal actual evapotranspiration and soil water uptake. The search for an appropriate canopy water potential is made by an iterative procedure. Rainfall and potential evapotranspiration are entered daily while the simulations are also made with a 24 hr time increment. A flow diagram and a detailed description of the model can be found in Rambal (1987). Good agreement between simulated and measured values of soil water content (Rambal 1987) and canopy water potential (Rambal 1984a) indicate that the model has a good predictive value.

Computed water yield for existing land-use (1979) corresponds with recorded water yield on both the monthly (Figure 6–7) and annual time scales. Thus, for the two consecutive years 1979 and 1980, the simulated annual water yields were 411 mm and 212 mm, respectively, and the measured ones were 447 mm and 161 mm, respectively.

Land-use changes, due to changes in vegetation from 1946 to 1979, show a slow replacement of the shrublands by the woodlands, which were reduced from 83% cover in 1946 to 22% cover in 1979 (Figure 6–8). During this period, disturbance, including clearcutting of coppices and some natural or controlled fires, was infrequent. Vegetation change during this period was dominated by the recovery processes (see Debussche et al. 1987). These changes in vegetation produced an 80-mm reduction in the mean annual water yield for an average rainfall of 1200 mm.

For 1946: $Q = 1.031 (P - 521)$ and for 1979: $Q = 0.920 (P - 558)$.

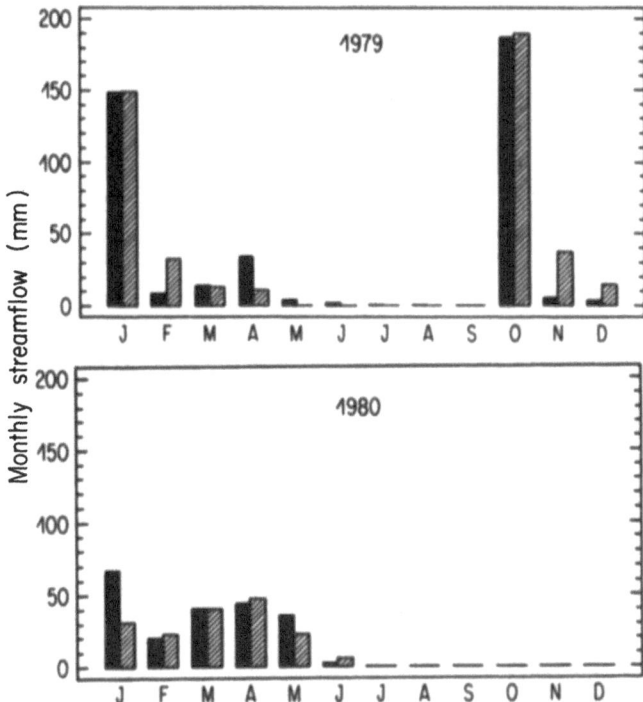

Figure 6–7. Comparison of measured (solid) and simulated (diagonal lines) monthly streamflow of a karstic watershed near Montpellier, France, for the years 1979 to 1980.

Advances in satellite remote-sensing technology in the last decade offer a new tool for measuring terrestrial vegetation cover and LAI. The combination of red and near-infrared reflectances has been found to be strongly related to canopy characteristics such as leaf area index. Data from a Spot satellite overpass on August 8, 1986, have shown a good relationship between the field measured LAI and the IR/Red combination (Lacaze 1990). So, remotely sensed LAI may provide a rapid and complete spatial coverage elimating the need for interpolation of plant. LAI when attempting to predict water use.

Water Balance of a *Quercus coccifera* Garrigue During the Regrowth Period Following Fire: Simulation and Some Related Consequences

The study site is located 10 km north of Montpellier, France, at the top of a west-facing 15% slope. The site is heterogeneous and comprises soft to

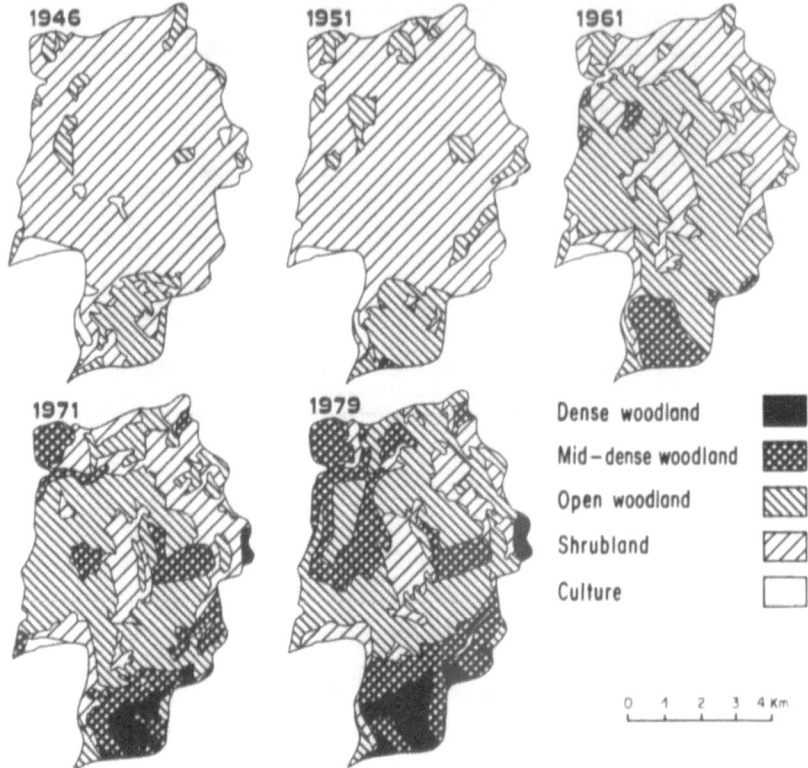

Figure 6–8. Change in the land use of the karstic watershed from 1946 to 1979.

hard limestone covered with a very shallow soil mantle. Clay loam soil fills the cracks and fractures. The vegetation was a dense continuous canopy of shrubs unburned since 1951, and 95% of the shrub cover was provided by the broadly distributed *Quercus coccifera* L. The total leaf area index, averaged over the year, was about 2.4, and the rooting depth reached 4.7 m (see a more detailed description of vegetation and climate in Rambal 1984b). The model (Rambal 1988, 1990), was used to simulate water balance in 1980, a "normal" rainfall year with 636 mm (Rambal 1984b).

For the simulation year, the actual evaporation reached 577 mm and water yield or deep drainage was 161 mm. Of the 577 mm of water loss, 462 mm were by transpiration from the shrubs and 115 mm by evaporation from the soil. This resulted in a transpiration/actual evaporation ratio of 0.8 and a transpiration per unit of leaf area of 193 mm.

As indicated above, changing leaf area index (LAI) strongly modified the annual water balance. LAI directly controls the fractional absorption of radiation by the canopy and is, therefore, a driving variable in the initial routing of energy to the plant canopy and the soil surface. As leaf area indices increase, evaporation and water yield decrease, while transpiration increases (Figure 6–9). For a bare soil occuring immediately after a fire (LAI = 0), water yield reached 435 mm. At low leaf area indices (LAI <1), for instance during the first stage of regrowth after fire, the water yield sharply dropped. As the leaf area index increased from 1 to 4, water yield remained nearly constant and always lower than 150 mm. For LAI = 0, soil evaporation was 207 mm. Miller (1981) simulated an evaporation of 200 to 300 mm from bare soil in southern California and about 200 mm in central Chile. As the leaf area index increased from 0 to 4, soil evaporation was reduced to approximately 80%. Actual evaporation computed by adding soil evaporation and transpiration quickly increased when leaf area was increased from 0 to 2, showing that an important consequence of the reduction of the leaf area was the reduction of water loss, particurlarly at low leaf area indices. At high leaf area indices (LAI >2), actual evaporation increased to a plateau because increased transpirational losses were compensated for by decreased soil evaporation.

Poole and Miller (1981) observed that the boundary of chaparral and coastal sage occured where actual transpiration per unit leaf area index (T/LAI) reached about 200 mm a^{-1}. Poole and Miller (1981) gave similar annual transpiration rates in mixed and chamise (*Adenostoma fasciculatum* H. and A.) chaparral. Consequently, they assumed that as chaparral regrows after fire, the leaf area index increases and the transpiration per unit leaf area index decreases until T/LAI is reduced to 200 mm a^{-1}. At this value, Miller (1981), using field measurements and theoretical

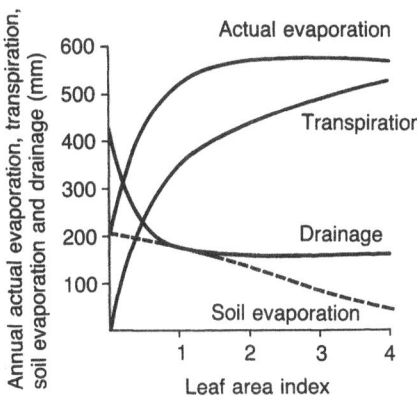

Figure 6–9. Annual actual evaporation, transpiration, evaporation from soil, and deep drainage (all in mm a^{-1}) with increasing leaf area index of a *Quercus coccifera* garrigue.

considerations on the carbon balance, concluded that the annual carbon fixed in photosynthesis is completely used in respiration to maintain the aboveground and belowground biomass. For $Q.$ $coccifera$, T/LAI was 193 mm a^{-1}, a value similar to the one proposed above. Hence, we also adopt 200 mm a^{-1} as a threshold value necessary for this plant to maintain a positive carbon balance.

Conjectures About the Possible Effects of "Global Change"

To determine the effects of global changes on the water balance, the summer plant water status and, consequently, the flammability of a $Quercus$ $coccifera$ garrigue, scenarios involving changes in temperature, precipitation, and plant responses to the increased atmospheric concentration of carbon dioxide was used to drive our canopy model. Hypothetical climate-change and plant responses were chosen for analysis.

General Circulation Models (GCM), in the most conservative simulations (taking into account the feedback effect of clouds on the energy budget of the earth), forecast for the southern part of Europe under a fourfold-higher CO_2 level ($4 \times CO_2$) atmosphere, a 15-day advance in soil drying (see, for instance, Manabe and Wetherald 1987). At a scale more appropriate to define regional scenarios of climatic change, Wilson and Mitchell (1987) simulated for the southern part of France a $4 \times CO_2$ decrease in daily rainfall of 1 mm and of 0.5 mm during the winter and summer months, respectively. Their scenario for air temperature suggest an increase $+4°C$ in January and $+7°C$ in July (Figure 6–10). While this estimate is rather severe, the general consensus is that doubling of atmospheric carbon dioxide could lead to an average global warming of between 1.5 and 4.5°C, with a most likely temperature increase of approximately 3.0°C (Manabe and Stouffer 1980; Hansen et al. 1983; Washington and Meehl 1983). Wide variations in precipitation are also expected. For the purpose of this study, I evaluated the effects of decreasing precipitation by 20%. The remaining analysis was based on plant response to increases in CO_2. Concerning the physiological effects of CO_2, we used the experimental results of Tenhunen et al. (1989), obtained with Mediterranean sclerophyllous $Quercus$ $spp.$, including $Q.$ $coccifera$. This study, showed that leaf stomatal conductance was reduced by 40% under $2 \times CO_2$. This decrease seemed to be independent of the plant water status. Following Shugart et al. (1985), who wrote that "the limited evidence available suggest that growth in height, leaf area, and dry weight of temperate trees is increased by higher than normal CO_2 concentration", we simulated a hypothetical 50% increase in leaf area index.

All these scenarios can be summarized as follows: (1) a scenario (S1) of temperature increase ($+3°C$) and current precipitation and another scenario involving a combination of $+3°C$ and $-20%$ precipitation (S2);

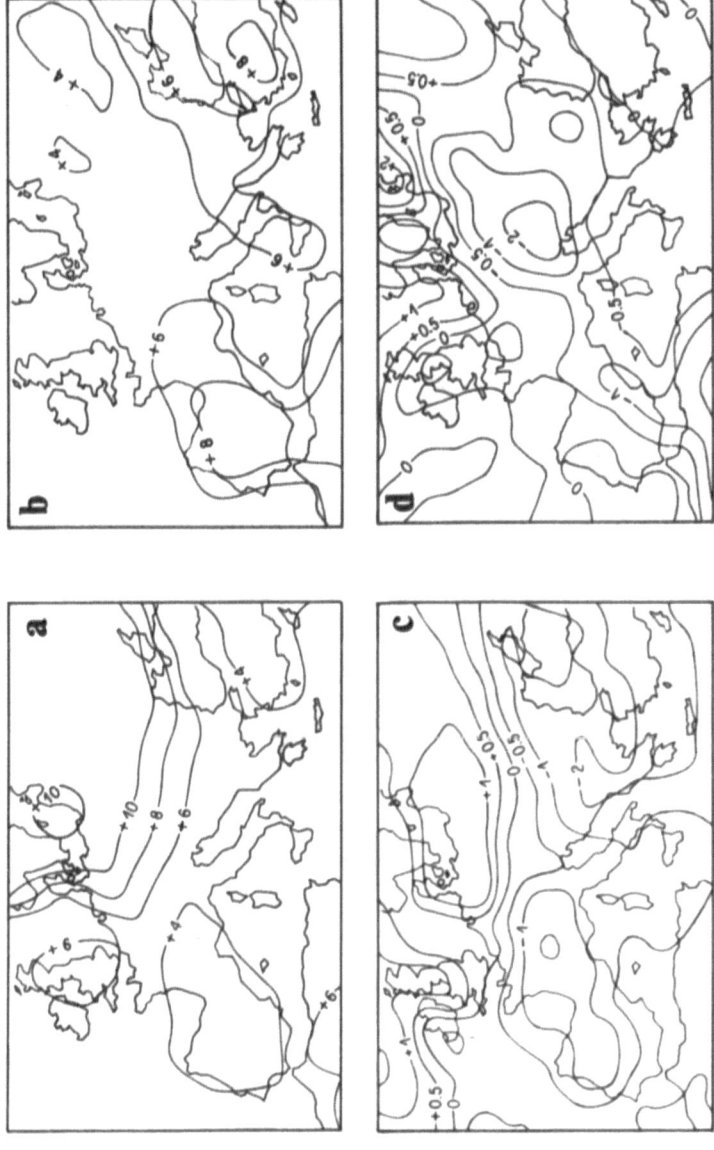

Figure 6-10. Change in January and July mean temperatures (**a** and **b**) and mean daily rainfall (**c** and **d**) under a $4 \times CO_2$ hypothesis, using the UK Meteorological Office 5-layer General Circulation Model. (Redrawn from Wilson and Mitchell 1987.)

(2) a scenario (S3) of stomatal canopy conductance (−40%) and another
involving a combination of −40% conductance and +50% LAI (S4); (3)
a scenario (S5) involving a combination of +3°C and +50% LAI and
−20% precipitation and −40% conductance.

We chose two consecutive simulation years, 1980 and 1981; "normal
years" that received 636 and 673 mm of annual rainfall, respectively.
These values produced very similar results of simulated actual evapotranspiration (576 and 573 mm respectively), of which evaporation from
the soil represented 20 and 21% respectively, and yielded very con-
strasted deep drainage (161 mm in 1980 and 19 mm in 1981). The difference is explained by large variations in the soil water storage. The
summer stress indexes or the leaf water potential averaged over the
period of July through September reached −10.1 and −9.4 bars, respectively. These results, with those of the 5 scenarios, are shown in
Figure 6–11.

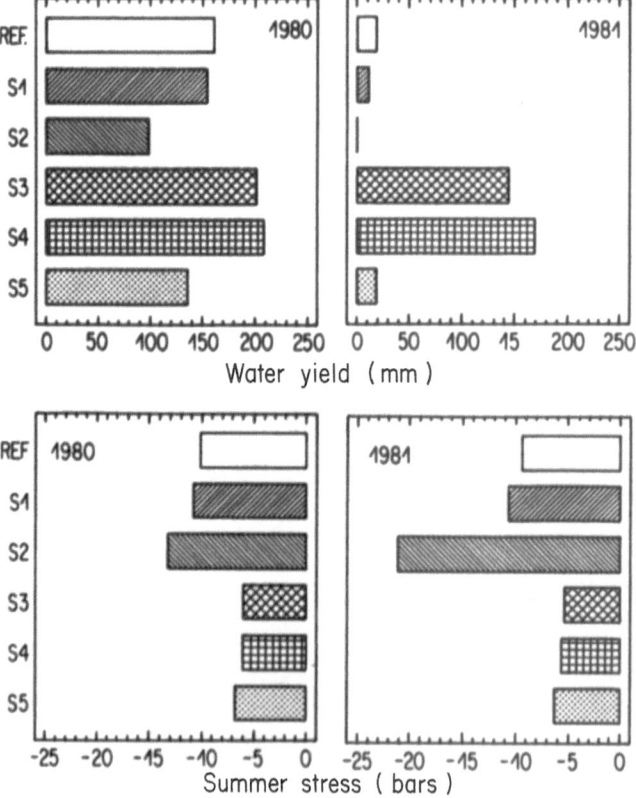

Figure 6–11. Comparison of annual steamflow and the summer stress of the
reference case in 1980 and 1981 with those obtained in the case of five scenarios
(see text).

Increasing temperature (scenario S1) induced a small increase in actual evaporations (591 and 588 mm respectively), and consequently, a small decrease in water yield relative to the reference values (154 mm and 11 mm respectively). The summer stress indexes were −10.9 and −10.7 bars, respectively. The combination of +3°C and −20% precipitation (scenario S2) caused important changes in the water balance and in the plant water status. Actual evapotranspirations was 558 mm and 482 mm for 1980 and 1981, respectively. Soil evaporation reached 22% and 25% in these two years. The water yield sharply decreased to 97 mm in 1980 and nearly reached zero (0.2 mm) in 1981. The summer stress indices were −13.3 and −21.2 bars, showing increases of 32 and 126% relative to the reference values. Consequently, we can postulate an increase in the risk of fire due to the dryness of the fuel. In the case where only the doubling of atmospheric carbon dioxide was considered (S3), the actual evapotranspiration was 465 mm and 460 mm for 1980 and 1981, respectively, with the same percentage (25%) of total evaporation and soil evaporation. The reduction of the transpiration favored an increase in water yield. This reached 202 mm and 145 mm, in the 1980 and 1981 simulation, which corresponded to increases of 25% and 781%, compared to the reference values. The summer stress indexes remained still rather low, −6.1 and −5.7 bars, respectively.

In the following scenario (S4), the 50% increase in the leaf area index promoted a large extinction of the solar radiation through the canopy. Thus, soil evaporation levels were only 58 mm and 56 mm. Consequently, the rate of transpiration showed a slight increase of 12% relative to scenario S3. Deep drainage was 209 and 169 mm in 1980 and 1981, respectively. In the last scenario (S5), the water consumption of the soil–plant system (460 and 454 mm), and the summer stress indexes (−6.8 and −6.4 bars), were still less than the reference values. Water yields (136 mm and 19 mm respectively), were very similar to the reference values (161 mm and 19 mm respectively).

The main conclusion from these simulations was that a *Quercus coccifera* canopy under scenario S5 can maintain a higher leaf area than the reference case without undergoing greater water stress. The direct consequence of this result was that a greater leaf area index was possible. Nevertheless, it is necessary to discuss this outcome with respect to survival and long-term persistence of the plant. In the reference case, we measured water balance during 7 consecutive years (Rambal 1984a). From these data, we fitted the following linear regression between water yield and annual rainfall: $Q = 0.907 (P − 578)$.

This result indicates that when annual rainfall is less than 578 mm, deep drainage loss was negligible and almost all precipitation infiltrating into the soil is lost by actual evapotranspiration. Above this threshold the canopy consumed a small amount (9.3%) of the additional precipitation. For instance, with an annual rainfall of 868 mm (i.e., 300 mm above the threshold value) the actual evapotranspiration was only 578 + 0.093 × 300 = 596 mm. The water tapping system of *Q. coccifera* is rigid and adjusted to a low annual rainfall. The incomplete annual rainfall record of Montpellier since 1766 was adjusted to a lognormal distribution with a mean of 761 mm and a standard deviation of 223 mm. Annual rainfall of 578 mm has an occurrence probability of 0.116, that is a return period of 8.6 years. If the mean rainfall is decreased by 20%, the occurrence frequency of the 578 mm annual rainfall changes drastically to 0.388, corresponding to a return period of 2.6 years. Under these circumstances, an event occurring once every century will now happen every 9.1 years.

Evaluating the consequences of global changes effects on a perennial species ecosystem should include an evaluation of the new functional equilibrium of LAI. One must take into account the possibilities offered to the plant by increased atmospheric CO_2 as well as the increased probability of drought if decreased annual rainfalls are predicted.

Acknowledgment. The comments of W.C. Oechel and A. Finzy are gratefully acknowledged.

References

Anderson, H.W. 1949. Does burning increase surface runoff? *J. Forestry*, 47: 54–57.

Bosch, J.M., Hewlett, J.D. 1982. A review of catchment experiments to determine the effect of vegetation changes on water yield and evapotranspiration. *J. Hydrol.* 55:3–23.

Bosch, J.M., Van Wilgen, B.W., Bands, D.P. 1986. A model for comparing water yield from fynbos catchments burnt at different intervals. *Water SA* 12:191–196.

Brown, T.C., O'Connell, P.F., Hibbert, A.R. 1974. Chaparral conversion potential in Arizona. Part II: An economic analysis. USDA For. Serv. Res. Pap. RM-127, Rocky Mt. For. and Range Exp. Stn., Fort Collins, CO.

Daniell, T.M., Kulik, V. 1987. Bushfire hydrology. The case of leaking watersheds. *J. Hydrol.* 92:301–313.

Debussche, M., Rambal, S., Lepart, J. 1987. Les changements de l'occupation des terres en région méditerranéenne humide: évaluation des conséquences hydrologiques. *Acta Oecol., Oecol. Applic.* 8:317–332 + maps.

Hamilton, L.S. 1988. Forestry and watershed management, pp. 99–111. *In* J. Ives and D.C. Pitt (eds.), *Deforestation: Social Dynamics in Watersheds and Mountain Ecosystems*. Routledge, London, England.

Hansen, J., Russell, G., Rind, D., Stone, P., Lacis, A., Lebedeff, S., Ruedy, R., Travis, L. 1983. Efficient three-dimensional global models for climate studies: Models I and II. Monthly Weather Review 111:609–662.

Hibbert, A.R. 1967. Forest treatment effects on water yield, pp. 527–543. *In* W.E. Sopper and H.W. Lull (eds.), *Forest Hydrology*, Pergamon, Oxford, England.

Hibbert, A.R., Davis, E.A., Scholl, D.G. 1974. Chaparral conversion potential in Arizona. Part II: Water yield response and effects on other resources. USDA For. Serv. Res. Pap, RM–126, Rocky Mt. For. and Range Exp. Stn., Fort Collins, CO.

Holmes, J.W., Wronski, E.B. 1982. On the water harvest from afforested catchments. *Proceedings of the First National Symposium on Forest Hydrology*, Melbourne, Australia.

Hoyt, W.G., Troxell, W.C. 1932. Forest and streamflow. *Proc. Am. Soc. Civ. Eng.* 56:1037–1066.

Kuczera, G. 1987. Prediction of yield reductions following a bushfire in ash-mixed species Eucalyptus forest. *J. Hydrol.* 94:215–235.

Lacaze, B. 1990. The relationships between remotely sensed vegetation indices and plant canopy properties, pp. 137–151. *In* E.C. Barrett, C.H. Power, and A. Micaleff (eds.), *Satellite Remote Sensing for Hydrology and Water Management* Gordon and Breach Pub, Montreux, Switzerland.

Langford, K.J. 1976. Change in yield of water following a bushfire in a forest of *Eucalyptus regnans. J. Hydrol.* 29:87–114.

Le Houérou, H.N. 1987. Vegetation wildfires in the Mediterreanean basin: Evolution and trends. *Ecologia Mediterranea* 13:13–24.

Lieutaghi, P. 1972. *L'Environnement Végétal. Flore, Végétation et Civilisation.* Delachaux et Niestlé, Neuchatel, Suisse.

Manabe, S., Stouffer, R.J. 1980. Sensitivity of a global climate model to an increase of CO_2 concentration in the atmosphere. *J. Geophys. Res.* 85: 5529–5554.

Manabe, S., Wetherald, R.T. 1987. Large-scale changes of soil wetness induced by an increase in atmospheric carbon dioxide. *J. Atmospheric Sci.* 44: 1211–1235.

McNaughton, K.G., Jarvis, P.G. 1983. Predicting effects of vegetation changes on transpiration and evaporation, pp. 1–47. *In Water deficits and plant growth*, (Vol. 7). Academic Press, London.

Miller, P.C. 1981. Similarities and limitations of resource utilization in Mediterranean-type ecosystems, pp. 369–407. *In* P.C. Miller (ed.), *Resource Use by Chaparral and Matorral. A Comparison of Vegetation Function in Two Mediterranean-Type Ecosystems.* Ecological Studies, 39. Springer-Verlag, Berlin.

Poole, D.K., Miller, P.C. 1975. Water relations of selected species of chaparal and coastal sage communities. *Ecology* 56:1118–1128.

Poole, D.K., Miller, P.C. 1981. The distribution of plant waterstress and vegetation characteristics in southern California chaparral. *Amer. Midl. Nat.* 105:32–43

Rambal, S. 1984a. Sécheresse réelle, sécheresse calculée. *Bull. Soc. Bot. Fr. Actual. Bot.* 131:295–301.

Rambal, S. 1984b. Water balance and pattern of root water uptake by *Quercus coccifera* L. evergreen scrub. *Oecologia* 62:18–25.

Rambal, S. 1987. Evolution de l'occupation des terres et ressources en eau en région méditerranéenne karstique. *J. Hydrol.* 93:339–357.

Rambal, S. 1988. A simulation model for predicting water balance and canopy water potential of a *Quercus coccifera* garrigue. *Ecologia Mediterranea* 14:95–99.

Rambal, S. 1990. From daily transpiration to seasonal water balance: An optimal use of water? *In* J. Roy and F. di Castri (eds.), *Time Scales of Biological Responses to Water Constraints.* Springer-Verlag, New York.

Rambal, S., Lacaze, B., Mazurek, H., Debussche, G. 1985. Comparison of hydrologically simulated and remotely sensed actual evapotranspiration from some Mediterranean vegetation formations. *Int. J. Remote Sensing* 6: 1475–1481.

Shachori, A.Y., Michaeli, A. 1965. Water yields of forest, maquis and grass covers in semi-arid regions: A literature review, pp. 467–477. *In* F.D.

Eckardt (ed.), *Méthodologie de l'Écophysiologie Végétale*. UNESCO, Paris, France.

Shugart, H.H., Antonovsky, M.Y.A., Jarvis, P.G., Sandford, A.P. 1985. CO_2, climatic change and forest ecosystems. Assessing the response of global forests to the direct effects of increasing CO_2 and climate change, pp. 475–521. *In* B. Bolin, B.R. Döös, J. Jäger, and R.A. Warrick (eds.), *The Greenhouse Effect, Climatic Change and Ecosystems*. Scope 29. Wiley and Sons, Chichester, England.

Tenhunen, J.D., Reynolds, J.F., Lange, O.L., Dougherty, R.L., Harley, P.C., Kummerow, J., Rambal, S. 1989. Quinta: A physiologically-based growth simulator for drought adapted woody plant species, pp. 135–168. *In* J.S. Pereira and J.J. Landsberg (eds.), *Biomass Production by Fast-Growing Trees*. Kluwer Academic Publishers.

Trabaud, L. 1987. Natural and prescribed fire: survival strategies of plants and equilibrium in Mediterranean ecosystems, pp. 607–621. *In* J.D. Tenhunen, F.M. Catarino, O.L. Lange, and W.C. Oechel (eds.), *Plant Response to Stress. Functional Analysis in Mediterranean Ecosystems*. NATO Advanced Science Institute Series, Vol. G15, Springer-Verlag, Berlin, Germany.

Trimble, G.R. Jr., Reinhart, K.G., Webster, H.H. 1963. Cutting the forest to increase water yields. *J. Forest.* 61:635–640.

Washington, W.M., Meehl, G.A. 1983. General circulation model experiments on the climatic effects due to a doubling and quadrupling of carbon dioxide concentration. *J. Geophys. Res.* 88:6600–6610.

Wilson, C.A., Mitchell, J.F.B. 1987. Simulated climate and CO_2-induced climate change over western Europe. *Climate Change* 10:11–42.

7. Spatial Simulation of Fire Regime in Mediterranean-Climate Landscapes

Frank W. Davis and David A. Burrows

The kind of fire history that characterizes an area can be termed its fire regime, the elements of which are fire type and intensity, size, return interval, and spatial pattern. Fire regime plays a major role in determining regional patterns of species distributions, vegetation patterns and fluxes of matter and energy (Kilgore 1981; Johnson 1979; Knight 1987).

Although research in fire ecology has concentrated on succession after individual fires, there is a recognized need for longer-term perspectives on the relationships between fire regime, landscape heterogeneity, and human activities (Heinselman 1973; Johnson and Van Wagner 1985; Baker 1989b; Keane et al. 1990; Turner and Romme 1990). This perspective is being developed through the combination of empirical studies of regional fire histories and simulation models based on probabilistic rules or on physically based equations of fire ignition and fire spread. Simulation modeling is especially valuable for studying the sensitivity of fire regime to human activities such as fragmentation of natural vegetation, or to changing environmental conditions such as predicted global climate change (Turner et al. 1989; Baker 1989a).

Most fire models have treated fire behavior and postfire succession at a point or within a plot, neglecting spatial variations in physical and biological conditions and the interactions between landscape heterogeneity, fire spread, and fire regime (Turner and Romme 1990). Recent work by landscape ecologists has shown that these spatial effects can be significant (e.g., Turner 1989), and efforts are underway to incorporate realistic spatial variation into fire spread models. Such modeling is now facilitated by digital satellite and terrain data that provide information on land surface conditions, and by Geographic Information System (GIS) hardware and software for spatial data acquisition, storage and analysis (Burrough 1986).

In this chapter, we review current approaches to modeling fire regime, and discuss some of the uses of digital geographic data for studying fire regime in heterogeneous Mediterranean-climate landscapes. We introduce the REFIRES (REgional FIre REgime Simulation) model, a GIS-based simulation model, and demonstrate its use for studying fire regime in a human-altered landscape in southern California. Effects of changing ignition frequency and vegetation fragmentation are explored. We conclude with a discussion of potential applications and limitations of REFIRES, and suggest ways that it could be improved to obtain more realistic results.

Fire Regime in Mediterranean-Climate Ecosystems

Fire plays a dominant role in determining the structure and the functional properties of natural vegetation in Mediterranean-climate regions, especially in sclerophyll shrubland and woodland vegetation types such as the chaparral of California (Sweeney 1956; Hanes 1971) and the maquis and garrigue ecosystems of the Mediterranean Region (e.g., Naveh and Whittaker 1979; Trabaud and Lepart 1980). In chaparral ecosystems, fire recurs every 20 to 100 years, its spread promoted by continuous shrub canopy, high proportion of fine fuel, and low fuel moisture during the prolonged summer drought period. Wildfires are typically extremely hot, removing all but the largest stems. Chaparral stands regenerate rapidly after fire from vegetative sprouts and from buried refractory seed (Horton and Kraebel 1955). Annual species dominate the vegetation during the first year or two after burning. Subshrubs and perennial herbs attain maximal cover 3 to 7 years after fire, followed by closure of the shrub canopy over the next 5 to 10 years (Hanes 1971). As the stand ages, its flammability increases due to accumulation of a standing fine-litter fraction.

Stand-replacing fires in Mediterranean shrublands produce a mosaic of even-aged patches of vegetation in various stages of recovery (Minnich 1983). In theory, the spatial properties of these fire mosaics are closely coupled to other elements of regional fire regime such as ignition frequency and timing, fire type and fire intensity, fire management, postfire vegetation recovery and land use patterns (Parsons 1976; Minnich 1983; Turner et al. 1989). Some of these relationships are illustrated in Figure 7–1. In areas with infrequent ignitions or active fire suppression, fuel accumulation fosters infrequent, large, hot burns. Areas with frequent

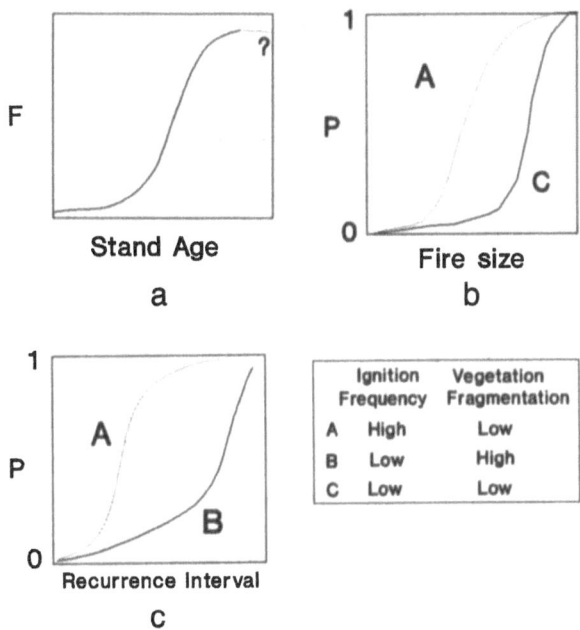

Figure 7–1. Some hypothesized relationships of fire regimes in chaparral ecosystems: (a) the relationship between stand flammability and stand age; (b) cumulative frequency distribution (P) of fire sizes for different ignition regimes and landscapes; (c) cumulative frequency distribution (P) of fire recurrence interval for different ignition regimes and landscapes.

ignitions and where fire is not suppressed should tend to develop vegetation mosaics with smaller patches and a greater diversity of stand ages because of the tendency for fires to extinguish at the boundaries of recent burns where there is insufficient fuel to carry the burn. Where vegetation has been fragmented by conversion of land to other vegetation types or land uses, fire spread is retarded and mosaic elements are decreased in size but may increase in average age.

Documenting Regional Fire Regime in Chaparral Ecosystems

Knowledge of the relationship between fire history, modern fire regime, and vegetation pattern is of theoretical and of practical significance for understanding chaparral dynamics on time scales of decades to centuries or millennia. Unfortunately, it is difficult to assemble a long fire history for shrubland ecosystems. Charcoal records in sediments have provided some information on prehistoric fire regimes. For example, Byrne et al. (1977) analyzed charcoal stratigraphy in varied sediments of the Santa

Barbara channel to reconstruct the prehistoric fire regime of the Santa Ynez Mountains. They concluded that the recurrence interval for wildfires in the Santa Ynez Mountains has remained at 60 to 100 years for the past several centuries, despite changes in fire management policies. Interpretations were limited by the inability to reliably distinguish the source area and fire size from fossil charcoal.

Historical narratives have been used to reconstruct fire regimes prior to the modern era of fire suppression and controlled burning. For example, Minnich (1987) reconstructed the behavior and distribution of several uncontrolled chaparral fires in southern California. One of the important conclusions from this study was that chaparral fires could continue to spread over periods of weeks and months, smoldering as embers in stumps and logs during cooler or more humid periods and flaring and running during hotter and windier periods.

Recent historic fire regimes can be studied using archival photography and, more recently, satellite imagery (e.g., Minnich 1983; Chuvieco and Congalton 1988; Davis et al. 1988). Interpretation of observed patterns is complicated by regional variation in factors such as climate and topography, and by historical changes in fire management policies. Nevertheless, fire history maps can help to identify determinants of fire regime, especially in Mediterranean shrublands, where stand-replacing fires recur frequently and are easily mapped. For example, using Landsat MSS imagery, Minnich (1983) mapped all fires greater than 20 ha in a large area encompassing coastal southern California and the northern Baja Peninsula. His analysis indicated that fires were smaller in Baja, where fires were not suppressed, than in adjacent areas of California, where suppression resulted in less frequent, larger burns.

Simulating Fire Behavior and Fire Regime

Both physical and probabilistic approaches have been used to model fire spread across land surfaces. Most physical models of fire behavior are based on the fire spread equations of Rothermel (1972), which predict fire reaction intensity (heat energy/fuel bed area/time) and rate of spread (distance/time) in homogeneous and continuous fuel beds as a function of fuel properties and meteorological conditions.

Meteorological variables include temperature, humidity, wind speed, and wind direction. Specific fuel parameters required by these models include amount of fuel per land area, fuel size distributions, height of the fuel bed, fuel heat content, particle density and moisture content. Most vegetation types, especially forests, do not meet the precondition of a continuous and uniform fuel bed, and the model has been applied most successfully to grasslands and to dense shrublands such as chaparral.

The Rothermel model has been elaborated by Albini (1976), who developed a formulation to describe characteristics of large fuel burnouts. Albini (1979) also developed a model to predict spotting distances from fire fronts. Other models have been developed to predict wind speed at midflame levels (Albini and Baughman 1979), wind-driven fire size, and fire shape (Anderson 1983; Baker 1983).

Models of fire regime must account for several elements beyond those of the fire behavior and fire spread models, including models of postfire succession and fuel accumulation, fire ignition frequency and timing, and climatic variation. Existing approaches range from purely statistical and nonspatial to physically based, spatially explicit models.

Spatially aggregated approaches (Turner and Romme 1990) include fitting Weibull or negative exponential distributions to age structure data to infer fire recurrence (Van Wagner 1978; Johnson and Van Wagner 1985). This approach has been applied successfully to forest ecosystems, but to our knowledge it has never been attempted for shrublands.

A more complex spatially aggregated approach couples fire regime with stand-level successional models to study interactions between fuel accumulation, fire behavior, vegetation recovery, and community composition (e.g., Van Wagtendonk 1985; Keane et al. 1990). Malanson (1984) used the physically based fire behavior equations of Rothermel (1972) to simulate stand composition as a function of fire intensity and fire regime in California coastal sage scrub.

Fire regime models that account for landscape heterogeneity have used cellular automata, in which fire spread across a gridded surface depends on the state of each cell and its immediate neighbors. For example, Turner et al. (1989) modeled disturbance as a percolation process on binary maps. They varied class proportions (e.g., flammable vs. nonflammable) and disturbance intensity (probability of disturbance spreading to a neighboring cell of the same type) to study the interactions between landscape pattern and disturbance processes. Green (1989) also used a cellular automaton to simulate fire spread, seed dispersal, and interspecific competition in forest ecosystems. Fires were described by randomly located ellipses whose frequency and size were described by the Poisson and negative exponential distributions, respectively. Several different fuel models were used ranging from uniformly distributed with no regeneration to mixed fuel types with regeneration.

A desirable extension of the modeling efforts described above is to incorporate the biological and physical realism of the stand-based models with spatial and temporal heterogeneity of the more probabilistic approaches. Such models could be extremely useful for understanding and managing fire dynamics in specific regions. Fire spread equations have been used to simulate individual fires over gridded surfaces with varying fuel characteristics (Frandsen and Andrews 1979), but not to model longer-term fire regime.

Toward this goal, we developed REFIRES to simulate fire regime in actual landscapes that are represented by digital terrain data in a Geographic Information System (GIS) framework. Before describing the REFIRES model, we provide a brief overview of spatial data and GIS applications for fire modeling and management.

Geographic Information Systems for Fire History Analysis

Many spatial variables influence fire regime, notably ignition sources, topography, soils, vegetation and land use patterns. These data can be represented in a GIS in one of two fundamentally different data structures. Vector data structures represent spatial patterns using lines in continuous coordinate space (a point is a line with length zero), and are especially well suited to representing thematic maps (e.g., vegetation, soils), linear features (e.g., roads) and networks (e.g., drainage systems). Raster data structures divide space into fields or cells and provide information for each field. The most common raster structure in the square lattice whose values are stored as two dimensional arrays.

Grid raster structures are used for imaging systems and satellite scanners, and the uniform cell size and shape are well suited to spatial analysis and simulation modeling, such as the modeling with cellular automata that was described previously. However, the diagonal neighbors in the grid raster representation pose a problem in modeling contagious spread by fire, which is better handled by hexagonal tessellations.

GIS capabilities of surface modeling and map overlay are increasingly used for fire hazard mapping and for fire history analysis. A number of studies have demonstrated that satellite data can be used to map general fuel types (e.g., Sadar et al. 1982; Shasby et al. 1981; Yool et al. 1985). Cosentino et al. (1981) described an approach to fire hazard mapping in southern California that combined classified Landsat MSS data, digital topographic data, fire history maps, and site quality maps derived from soils, climate, and topographic data. As part of the same effort, Yool et al. (1985) tested the use of digital insolation maps to predict fuel loading in areas classified as chaparral, but obtained only modest agreement (37% variance explained). Using a similar approach, Burgan and Shasby (1984) combined classified Landsat MSS data with digital topographic data and meteorological data to maintain fire hazard maps based on the US National Fire Danger Rating System (NFDRS) (Deeming et al. 1977). Lu et al. (1990) described a microcomputer-based GIS that included physical, historical, and human activity data and used NFDRS equations to model several components of fire behavior and fire hazard.

Chuvieco and Congalton (1989) incorporated vegetation data derived from Thematic Mapper satellite imagery, topographic data, and road maps to map fire hazard on the Mediterranean coast of Spain on the basis of vegetation, slope, aspect, roads and elevation. The map was not a good predictor of the behavior of a documented fire but was effective in predicting the probability of an area burning.

Other recent applications of GIS to fire regime analysis include analysis of vegetation recovery from fires of varying intensity (Jakubauskas et al. 1990) and statistical assessment of vegetation succession under different management strategies (Lowell and Astroth 1989).

In most of the examples cited earlier, satellite-derived vegetation maps, digital terrain data, and fire history maps are the primary mapped variables that are used to describe fire regime and to parameterize mathematical models of fire hazard. All of these studies have used raster data structures whose spatial resolution appears to have been determined by the resolution of available satellite and terrain data. Cosentino et al. (1981) noted that the 80 m resolution of Landsat MSS data was adequate to describe vegetation stands in southern California, which tended to be at least twice the area of the MSS pixels. The authors do not know of any formal analysis of scale-dependent variation in fuel distribution and fire hazard conditions over large areas. Davis et al. (1989) documented considerable microscale (meters) variation in the distribution of fuel that was correlated with maximum temperature during burning and postfire succession in chaparral. Analyses over larger areas would be extremely valuable in selecting data resolution for fire modeling and in interpreting and applying model results.

Regional Fire Regime Simulation

REgional FIre REgime Simulation Model (REFIRES), uses existing fire spread equations to provide reasonable estimates of spread rates under varying meteorological and site conditions (see Burrows 1987 for more complete documentation and for program source code). For the purposes of simulating fire histories, we have been concerned mainly that the model predicts realistic fire sizes and shapes for the range of conditions likely to occur in the study region. We adapted the following model objectives:

1. Fire behavior should be modeled using the physically-based fire spread equations of Rothermel (1972) and Albini (1976).
2. The model of fire spread must account for the ability of unsuppressed fires to burn for weeks or longer, through diurnal and nocturnal meteorological conditions.
3. Spatial variation in topography, vegetation pattern, and fire history must be modeled explicitly.
4. REFIRES should be operable over relatively large grids in order to describe variation in fuels and topography over large regions.

5. The model should be modular to the degree possible to facilitate modifications and additions, and should not depend on specific GIS software.

REFIRES is written in the C programming language. The model operates on a surface of hexagonal cells. Each cell is described by fuel and topographic variables. The spread of fires through this grid is controlled by cell characteristics, along with sampled meteorological conditions, according to the predictions of the fire spread model developed by Rothermel (1972) and modified by Albini (1976). Other modifications to the rate of spread model are based on the work of Anderson (1983). Fire history is simulated over a period of years to centuries, as specified by the user.

A hexagonal grid is used because this tesselation provides equal distance to all cell neighbors. Also, the directional arithmetic of the hexagonal addressing system is well suited to modeling contagious diffusion processes such as the spread of disturbances through landscapes. We have adapted the Generalized Balanced Ternary addressing system described by Gibson and Lenzmeier (1981) and Gibson and Lucas (1982).

The general model flow is depicted in Figure 7–2. The landscape is described by digital maps of topography (elevation, slope angle, slope aspect), potential vegetation (may include land use classes such as agriculture or residential, which do not carry wildfires) and actual vegetation (a function of potential vegetation and time since last burning). Meteoro-

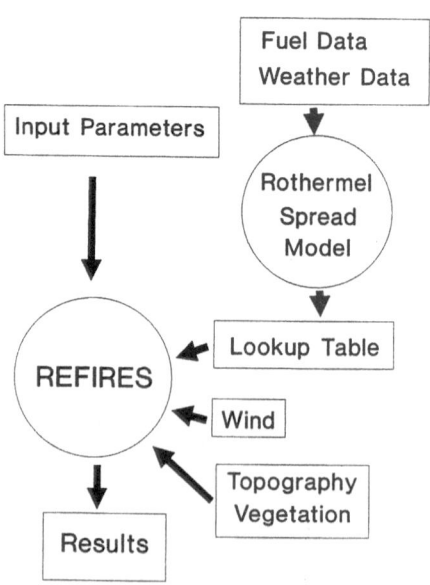

Figure 7–2. General formulation and data flow in the REFIRES model.

logical data consist of daily maximum and minimum temperatures and humidities, and accompanying live fuel moistures (by fuel size class) for a set of dates (we used a large random sample of fire season days from the study region). A frequency distribution of wind speeds and directions is also required.

At the beginning of a model run, the user inputs the grid size, the average number of possible ignitions per year (Fi), a threshold rate of spread (Rt, in feet per minute) below which fire spread does not occur, the number of years to run the model and a seed for a random number generator. The initial vegetation mosaic can be an actual pattern or any other vegetation map input by the user.

To reduce computation time, at the outset of a model run a lookup table is produced that contains the rates of spread (feet per minute) calculated by the Rothermel equations for each unique combination of vegetation (age classes within potential vegetation types), meteorological, and fuel moisture conditions. The parameters needed to build the lookup table are listed in Table 7–1. These spread rates are modified in REFIRES to account for wind and topographic effects.

For each year in the model, the number of possible ignitions is selected from a Poisson distribution with parameter Fi. The location of a possible ignition, date, time, and weather conditions are selected at random (Figure 7–3). The cell has a probability of ignition Pi that is calculated based on its dead fuel moisture and the weather conditions.

If the cell is ignited, the fire may spread throughout the cell based on the spread rate Rs as calculated from fuel, topography, temperature,

Table 7–1. Fuels and Weather Data Used To Calculate Rates of Fire Spread Under Conditions of No Wind and Level Topography

Fuels Data
Dead fuel surface area/volume (1/ft) by fuel size class
Live fuel surface area/volume (1/ft) by fuel size class
Heat content (BTU/lb)
Dead fuel load (tons/acre)
Live fuel load (tons/acre)
Fuel bed height (ft)
Dead fuel moisture of extinction (%)

Meteorological Data (entered by year, month, and day)
Day length (hours)
Maximum temperature (°F)
Minimum temperature (°F)
Maximum humidity (%)
Minimum humidity (%)

Wind Data (entered by wind direction, six classes for hexagon)
Relative frequency
Average speed

Note. Fuels data are entered by age class for each potential vegetation type in the region.

humidity and wind conditions. If Rs > Rt, then fire spreads through the cell with Ps = 1. If Rs < Rt, fire spreads through the cell with Ps proportional to Rs/Rt (see Figure 7–3). If fire spreads through the cell, the cell's age is set to 0 and fire then spreads to neighboring cells at rates that are calculated on a cell-by-cell basis based on local fuel conditions, topography, and wind conditions.

Using calculated rates of spread and distances, the program keeps track of the time elapsed between ignition and time when the burn reaches a cell. Daily maximum temperatures and minimum humidities are applied to cells that burn during daylight hours. Minimum temperatures and maximum humidities are applied during nighttime hours. At present, wind direction and velocity are kept constant over the length of a single burn. This is obviously too simplistic, and we intend to add a more realistic wind model. Also, fires spread only by contagious diffusion, not by spotting. A crude approximation to spotting can be implemented by setting minimum flammability to some value greater than 0, which allows fires to spread across fire breaks with low probability.

Fires cease spreading when they reach the edges of the grid or encounter conditions in which predicted spread rate falls below Rt, such as recent burns, nonvegetated areas, and areas with high fuel moisture content (e.g., riparian zones). Fires also may be extinguished due to changes in weather conditions or topography. After the model is run for

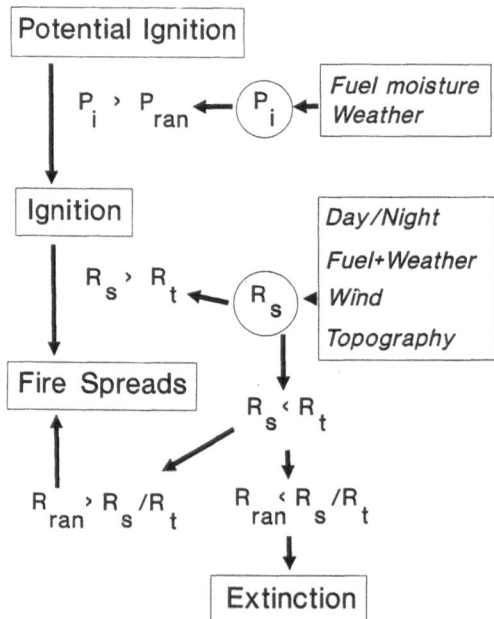

Figure 7–3. Inputs and operations for a fire event in REFIRES.

Table 7-2. Output of REFIRES

Run label
Grid size
Average ignitions per year (Pi)
Run length (yr)
Number of possible ignitions
Number of actual ignitions
Number of fires (by cell)
Fire size distribution
Age frequency distribution (overall and by type)
Patch size distribution
Fire recurrence interval (frequency distribution)

---

the specified number of years, the model outputs fire regime statistics and fire history maps (Table 7-2).

Model Application

We tested the general behavior of the model and sensitivity of some model parameters using data from Burton Mesa, near Lompoc in coastal southern California, 230 km northwest of Los Angeles (N 34° 42′ 30″, W 120° 30′, Figure 7-4). Climate is Mediterranean-type with a strong maritime influence and prevailing northwesterly winds. The topography is gently rolling uplands interrupted by wide stream valleys with short, steep slopes. The dominant vegetation is maritime chaparral, a shrubland type that is restricted to coastal locations on sandy soils. *Adenostoma fasciculatum* dominates mature stands (Davis et al. 1988). *Quercus agrifolia* is scattered throughout the chaparral, and forms closed forests on north-facing slopes and riparian environments. The native vegetation has been extensively fragmented by roads, residential areas, agriculture, and other developments (Figure 5). Although natural wildfire is rare, anthropogenic chaparral fires occur frequently (at least 27 times between 1938 and 1986). These fires are typically arrested at roads or fuels breaks, and thus give little indication of natural fire behavior (Figure 7-4).

We operated REFIRES over a 1865-ha region represented by 20,720 0.09 hectare cells. Minimum cell size was selected to capture the short forested slopes and riparian environments that could affect fire behavior. The study region was too small to model fire regime realistically over long time periods, because of the possibility of more remote ignitions spreading into the area. However, the small size made it easier to operate the model on a small computer (VAX 11/750) and to parameterize for initial tests.

The elevation grid was interpolated from the US Geological Survey 30-m Digital Elevation Model (DEM) for the Lompoc quadrangle. Two

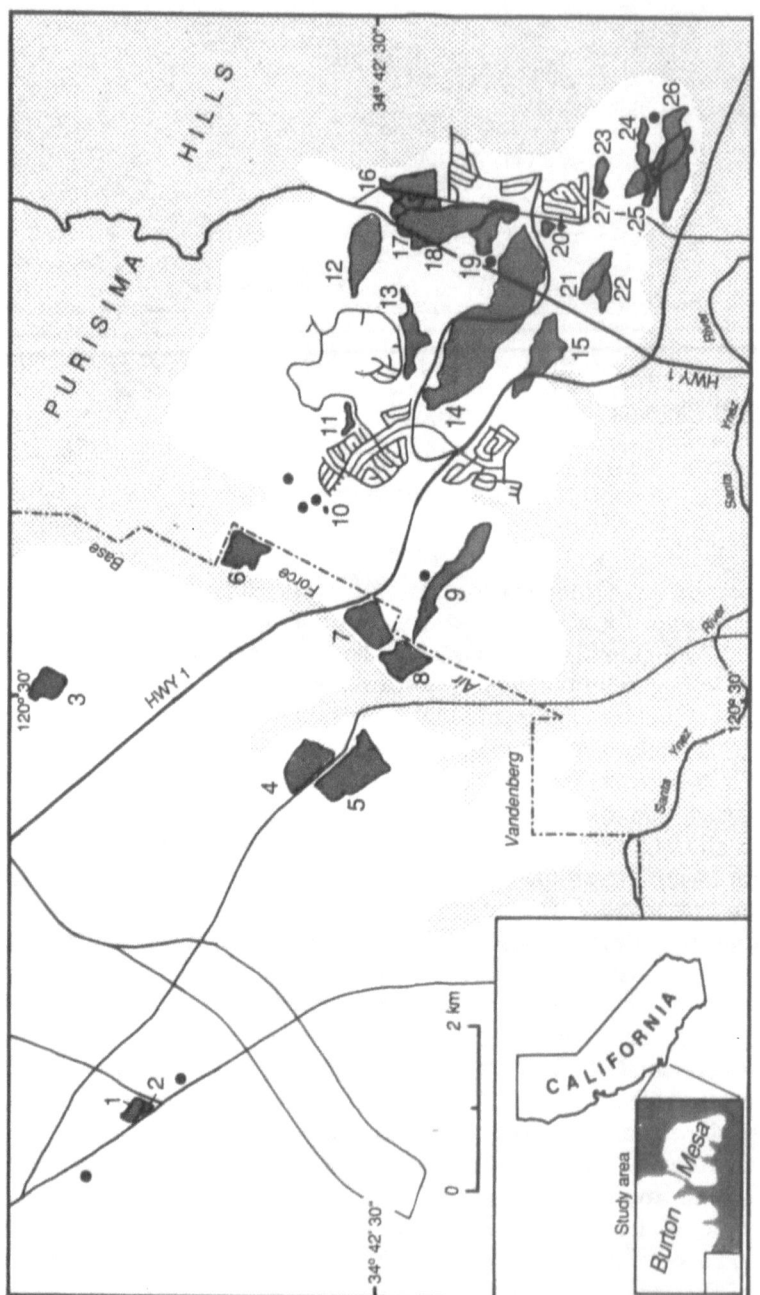

Figure 7–4. Map of study area showing Burton Mesa uplands (unshaded) and paved roads, as well as documented fires occurring between 1938 and 1985. (From Davis et al. 1988.)

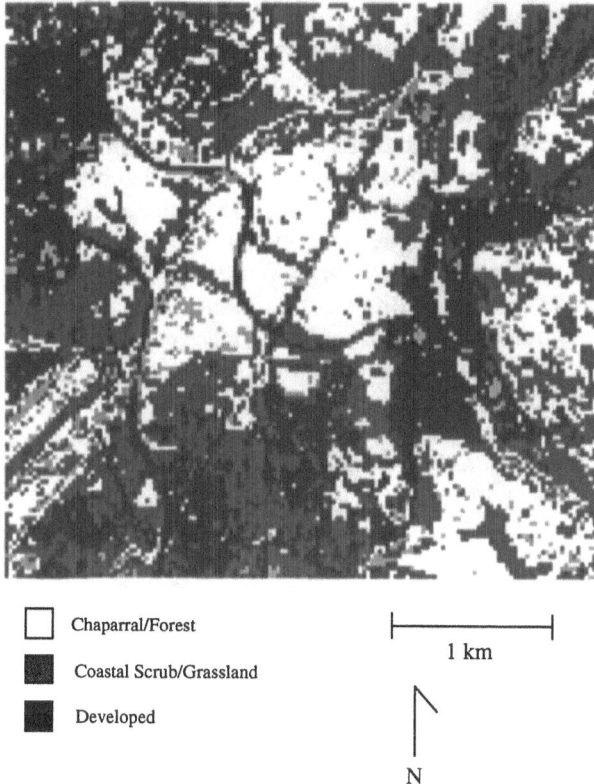

Chaparral/Forest

Coastal Scrub/Grassland

Developed

1 km

N

Figure 7–5. Land cover classification of the study region derived from 30 m Thematic Mapper Simulator data acquired July 1, 1984.

different potential vegetation models were used: (1) the entire area was mapped as *Adenostoma fasciculatum* chaparral, and (2) the region was modeled as chaparral, but areas mapped by Thematic Mapper satellite data as residential, agricultural or roads were mapped as nonflammable. We refer to these two vegetation models as Prehistoric and Modern.

The fuel model for *Adenostoma* chaparral was taken from the fuel load equations of Rothermel and Philpot (1973) and Bradshaw et al. (1983). Canopy heights for different age classes were taken from Rundel and Parsons (1979). Fire season (May through October) temperature and humidity data from nearby Santa Maria were obtained from the National Oceanographic and Atmospheric Administration. Corresponding green fuel moisture readings for *Adenostoma* stands were obtained from data collected 20 km to the east in the Los Padres National Forest (F. Cahill, unpublished data). Wind speed and direction data for Vandenberg Air Force Base, 10 km east of the study area, were obtained from Ogden (1975).

Ninety 500-year model runs (5 runs per parameter set) were conducted to test 18 different combinations of the following parameters: Prehistoric versus Modern vegetation; Fi = 0.1, 1.0 or 3.0 average possible ignitions per year; Rt = 1.0, 2.0 or 8.0 ft/min.

The Rothermel model has been tested, refined and validated over the past 15 years. However, we have not yet validated other aspects of REFIRES (such validation data must be obtained from a number of unsuppressed burns), and results reported here should be considered preliminary. Our main concern at this point is that model results appear reasonable and consistent with existing theory.

Ignition frequency and fire frequency increase linearly with increasing Fi (Table 7–3) on Prehistoric and Modern landscapes and for all values of Rt. The ratio of actual to possible ignitions is roughly 0.3 for the Prehistoric landscape and 0.2 for the Modern landscape, due to the fact that 34% of the Modern landscape is nonflammable. The ratio of ignitions to fires decreases with increasing ignition frequency, because with increasing ignition and fire frequency the average stand age decreases, reducing average flammability over the landscape.

Average fire size decreases nonlinearly with increasing ignition frequency under all scenarios except on the Modern grid with Rt set to 8 ft/min, when average fire size remains relatively constant. Fire size distribution on Prehistoric landscapes does not differ much between simulations using an Rt of 1.0 and 2.0 ft/min. At an ignition frequency of 0.1/yr, nearly half of the fires burned the entire landscape. The range of

Table 7–3. Statistical Summary of Simulated Fire Regime Under Varying Vegetation Fragmentation, Ignition Frequencies, and Threshold Rates of Spread

| | Vegetation Model | | | | | |
| | Prehistoric | | | Modern | | |
Ignition Frequency	0.1	1.0	3.0	0.1	1.0	3.0
Rt = 1.0						
Possible ignitions	49	495	1493	50	497	1496
Actual ignitions	14	155	464	10	90	307
Fires	11	83	226	9	64	176
Mean fire size	1052	285	123	268	137	70
Patch number	9	18	30	138	184	216
Rt = 8.0						
Possible ignitions	46	473	1471	49	90	303
Actual ignitions	13	135	452	9	90	303
Fires	9	55	176	6	56	157
Mean fire size	489	232	101	33	45	35
Patch number	16	28	47	28	211	314

Note. Values are means of five runs for each combination of modeled conditions.

fire sizes is much smaller on the Modern landscapes, and no fires spread over more than 50% of vegetated areas.

Figures 7–6 thru 7–9 show the frequency distributions of fire recurrence for different values of Fi and Rt on Modern and Prehistoric landscapes. These distributions are quite sensitive to changes in Fi, less so to Rt. For example, on the Prehistoric landscape with Fi = 0.1 and Rt = 1.0, 55% of burns spread across the entire landscape (see Figure 7–6), and 80% of

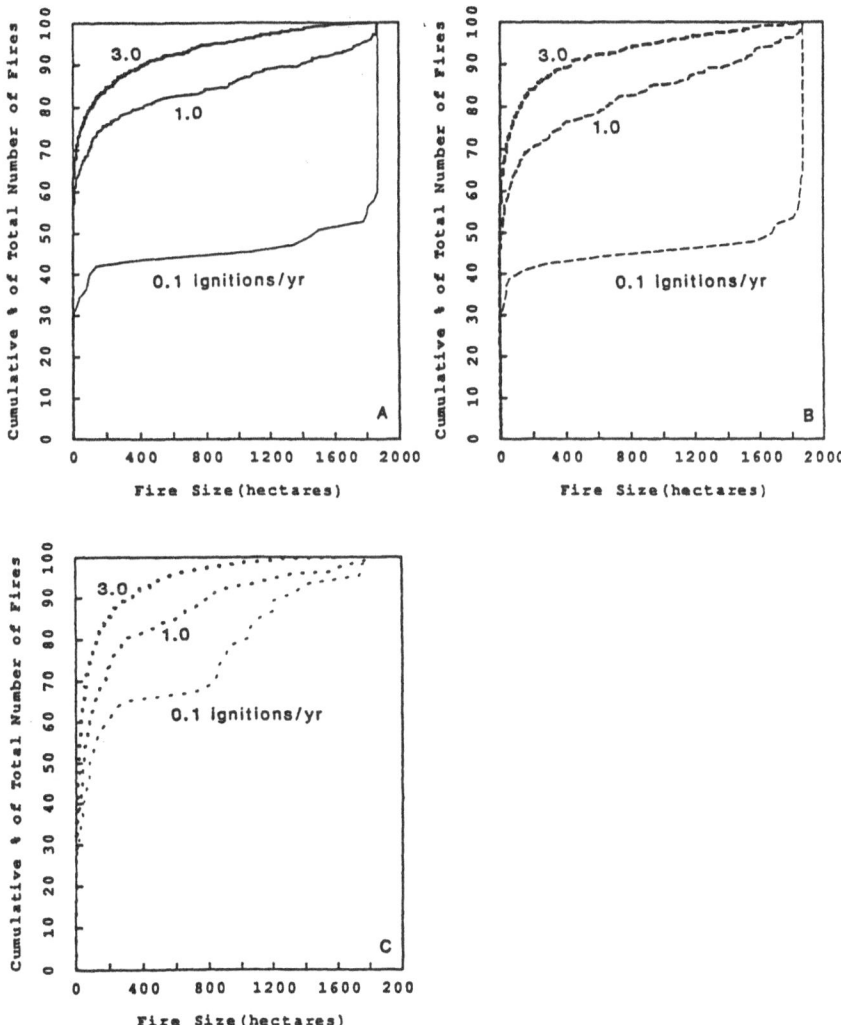

Figure 7–6. Cumulative fire size distributions obtained for Prehistoric landscapes with Fi = 0.1, 1.0 and 3.0 ignitions/yr, and for Rt = A, 1.0; B, 2.0; and C, 8.0 ft/min. Each parameter set was run five times to produce the curves.

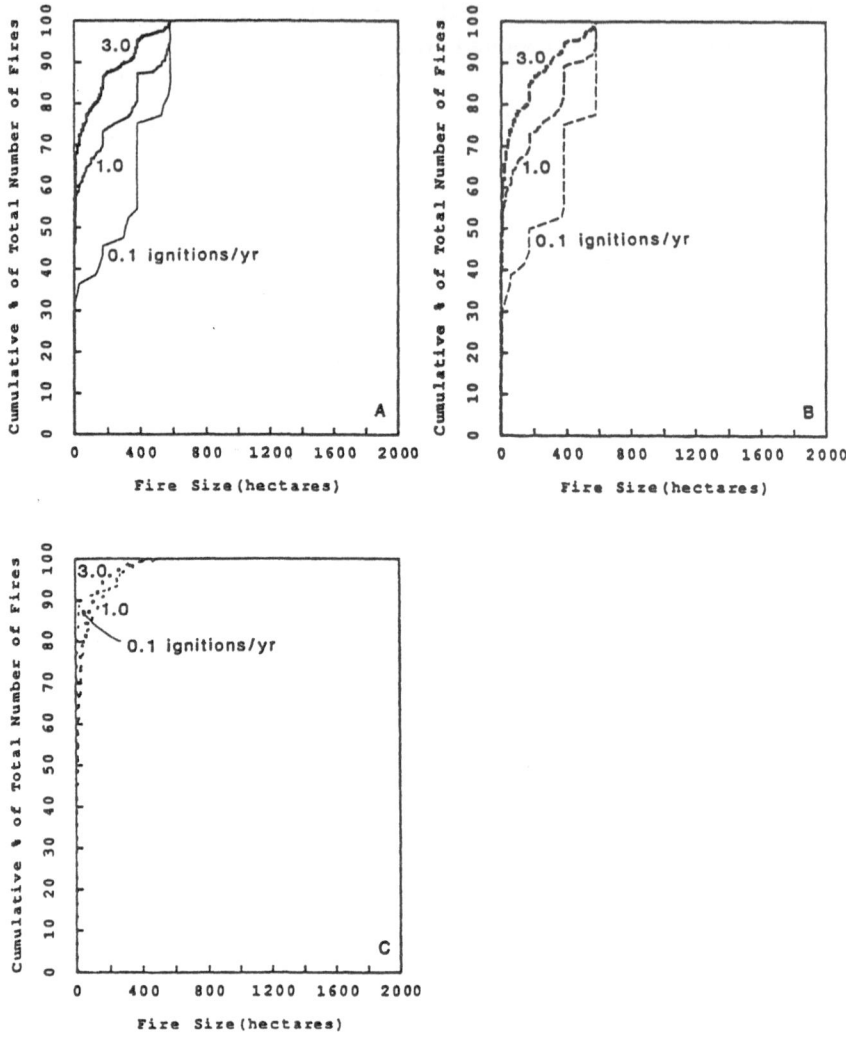

Figure 7–7. Cumulative fire size distributions for Modern landscapes for Fi = 0.1, 1.0 and 3.0 ignitions/yr, and for Rt = **A**, 1.0; **B**, 2.0; and **C**, 8.0 ft/min. Each parameter set was run five times to produce the curves.

cells burned at ages greater than 95 years (see Figure 7–8). With Fi = 1.0, roughly 75% of burns did not exceed 400 hectares in size, and 80% of the cells burned before reaching 45 years of age. Fire sizes are much smaller on the Modern landscape, and size distributions are less sensitive to Fi because of the isolation of old, highly flammable stands (see Figure 7–7). The fire recurrence distribution on the Modern grid is shifted strongly towards longer intervals (see Figure 7–9). With Fi = 0.1, a

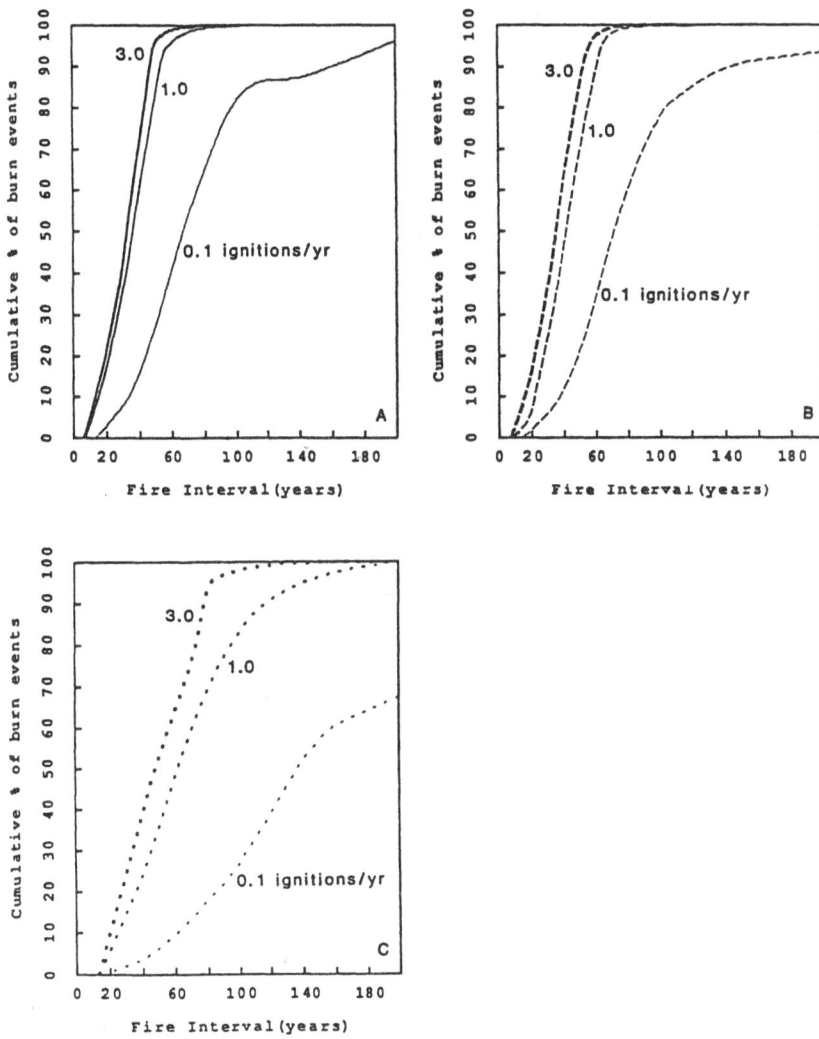

Figure 7–8. Cumulative frequency distribution of fire recurrence intervals obtained for Prehistoric landscapes with Fi = 0.1, 1.0, and 3.0 ignitions/yr, and for Rt = **A**, 1.0; **B**, 2.0; and **C**, 8.0 ft/min. Each parameter set was run five times to produce the curves.

cell has only 50% probability of burning during 200 years. Extensive agricultural and residential development has isolated many small chaparral patches that behave as so-called fire refugia. These areas are conspicuous in fire history maps, and, because of prevailing northwesterly winds, are especially common immediately to the east of extensive fuel breaks (Figure 7–10).

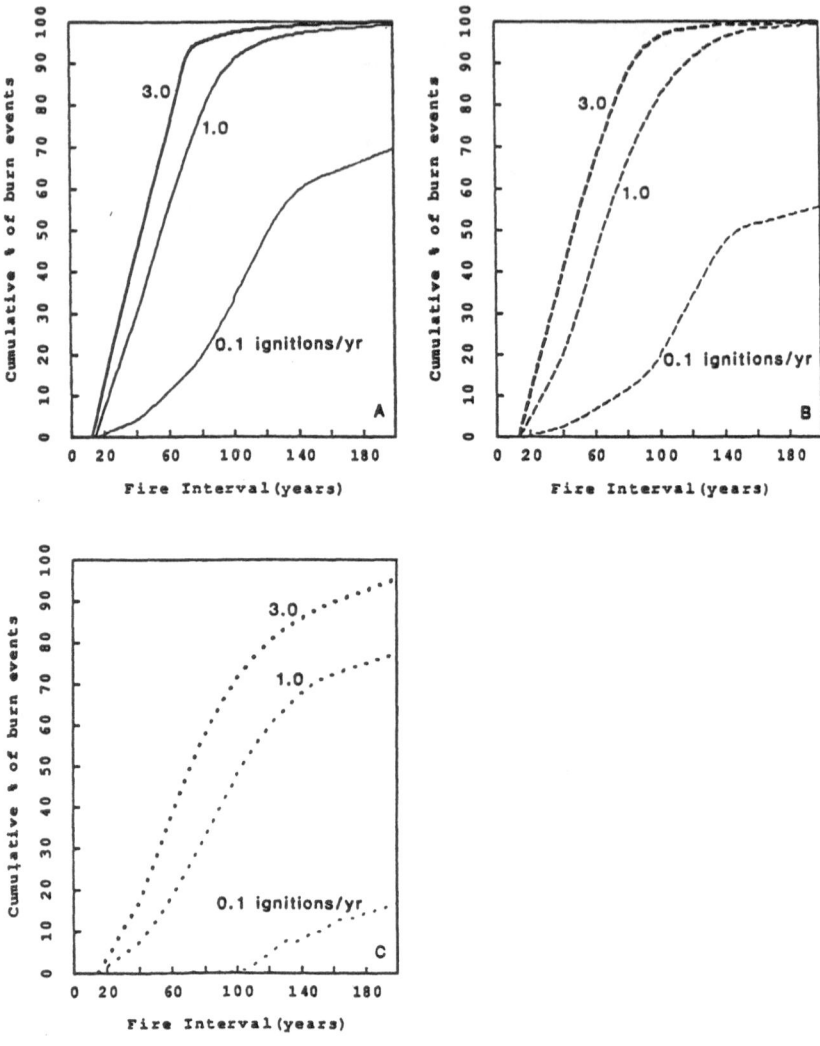

Figure 7–9. Cumulative frequency distribution of fire recurrence intervals obtained for Modern landscapes for Fi = 0.1, 1.0 and 3.0 ignitions/yr and for Rt = **A**, 1.0; **B**, 2.0; and **C**, 8.0 ft/min. Each parameter set was run five times to produce the curves.

Conclusion

Results obtained thus far with REFIRES appear reasonable and conform to the behavior of fire mosaics predicted by Parsons (1976), Minnich (1983), Turner et al. (1989) and others. Mosaic properties such as patch size and age distributions are closely coupled to ignition frequency, and

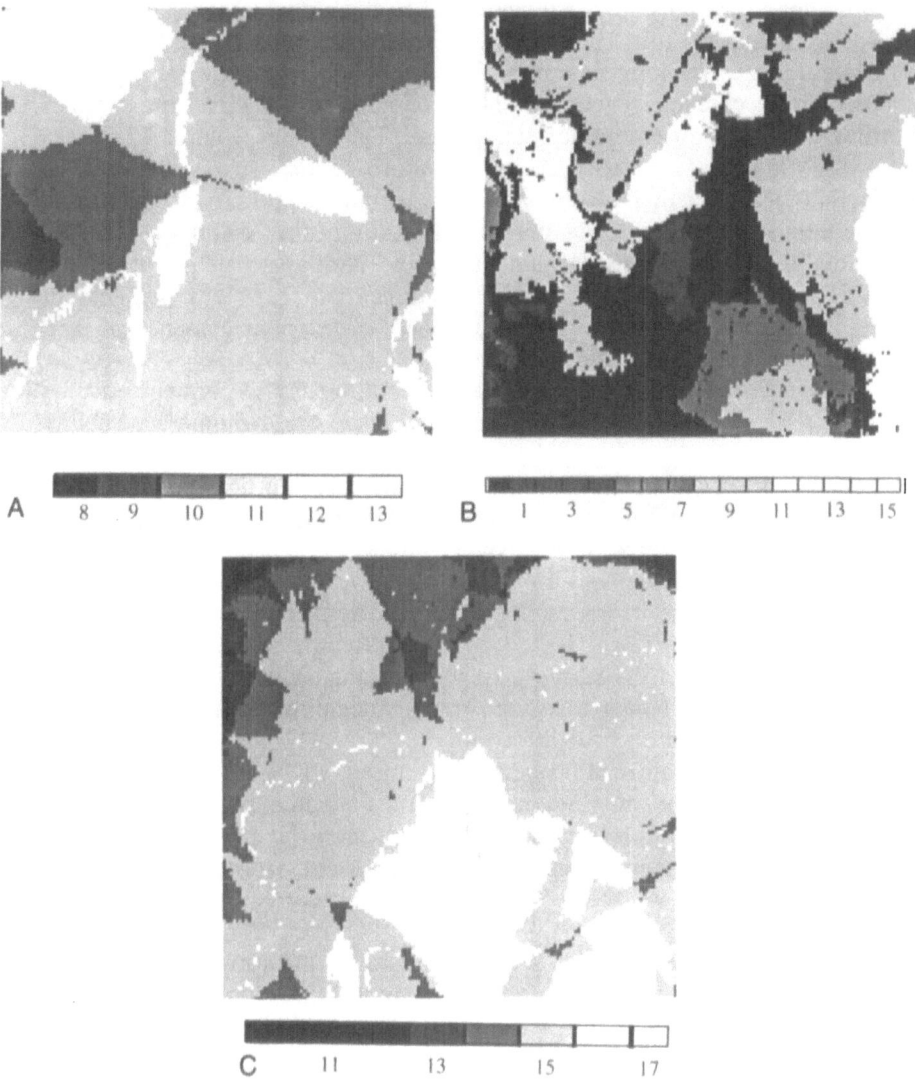

Figure 7-10. Fire histories produced by single 500-year runs of REFIRES. Image brightness is proportional to the number of fires. Rt was set to 1.0 ft/min for all runs. **A**, Prehistoric landscape, Pi = 0.1 ignition/yr; **B**, Prehistoric landscape, Pi = 1.0 ignitions/yr; **C**, Prehistoric landscape, Pi = 3 ignitions/yr; *D*, Modern landscape, Pi = 3 ignitions/yr.

are quite sensitive to vegetation fragmentation. The spatial distribution of fire recurrence is very uneven across the fragmented modern landscape, with fire refugia located, as expected, in areas generally downwind from large fuel breaks.

Despite the many stochastic elements to the model, there is only slight variation among runs made using the same parameters settings. This suggests that over long time periods (e.g., 500 years) and within a region of uniform climate, vegetation and physiography, chaparral fire mosaics may behave as quasi-equilibrium systems (Turner and Romme 1990).

One advantage of REFIRES over more general spatial fire models is that it uses actual terrain and climate data from a region. From a theoretical perspective, this permits the study of fire regime on realistic, spatially, and temporally dynamic surfaces. We hope that the model will ultimately be of practical use to resource managers and planners in indicating possible long term consequences of fire management and land use policies in fire-prone ecosystems.

Many of the terrain and the weather data needed to parameterize the model are readily available. Satellite data have the demonstrated ability to discriminate general fuel classes over large regions. The main obstacle is the lack of fuels information to parameterize the fire spread model; however, such data are also needed for other purposes by fire managers and thus are becoming increasingly available.

There are a number of improvements that can and should be made to REFIRES to make it more realistic. The current implementation locates ignitions at random in the region, but it is often the case that ignitions are strongly clustered in landscapes (e.g., lightning strikes near summits, or anthropogenic fires near roads and trails). Including a probability surface for ignition should be straightforward.

The meteorological model could be improved by introducing higher frequency variation in wind, temperature, and humidity. This could be accomplished by storing and accessing weather data chronologically. A fire that has ignited on a given day would progress through the daily sequence of weather conditions actually recorded for that year and season.

A model to simulate fire spotting should be incorporated. Equations have been developed to predict the probability, direction, and distance of spotting from burning trees (Albini 1979). A similar formulation for chaparral is recommended, but it introduces new complexity because of the need to simulate multiple fires burning simultaneously on the landscape.

One advantage of the current formulation is that it can be run on a relatively small computer. On the other hand, a large number of calculations are required to spread multiple fires across densely gridded surfaces (individual runs required several CPU hours on the VAX 11/750). We are currently investigating the linking of REFIRES to available GIS software, and the possibility of parallel processing to facilitate the analysis of larger grids and implementation of more complex formulations.

Acknowledgment. We would like to thank Jason Greenlee, Fritts Cahill of the Los Padres National Forest, and Bob Burgan for providing advice and information during model formulation. Development of REFIRES was funded by California Department of Parks and Recreation Contract 4252–609.

References

Albinet, G., Searby, G., Stauffer, D. 1986. Fire propagation in a 2–D random medium. *Le Journal de Physique* 47:1–7.

Albini, F.A. 1976. Estimating wildfire behavior and effects. *USDA Forest Service General Technical Report INT–30*. Inter-Mountain Forest and Range Experiment Station, Ogden, UT.

Albini, F.A. 1979. Spotfire distances from burning trees – a predictive model. *USDA Forest Service General Technical Report INT–268*. Inter-Mountain Forest and Range Experiment Station, Ogden, UT.

Albini, F.A., Baughman, R.G. 1979. Estimating windspeeds for predicting wildland fire behavior. *USDA Forest Service Research Paper INT–221*. Inter-Mountain Forest and Range Experiment Station, Ogden, UT.

Anderson, H.E. 1983. Predicting wind-driven fire size and shape. *USDA Forest Service Research Paper INT–305*. Inter-Mountain Forest and Range Experiment Station, Ogden, UT.

Baker, D.G. 1983. Shapes of simulated fires in discrete fuels. *Ecological Modeling* 20:21–32.

Baker, W.L. 1989a. Effect of scale and spatial heterogeneity on fire-interval distributions. *Canadian Journal of Forest Research* 19:700–706.

Baker, W.L. 1989b. Landscape ecology and nature reserve design in the Boundary Waters Canoe Area, Minnesota. *Ecology* 70:23–35.

Bradshaw, L.S., Deeming, J.E., Burgan, R.E., Cohen, J.D. 1983. The 1978 National Fire Danger Rating System: Technical documentation. *USDA Forest Service General Technical Report INT–169*. Inter-Mountain Forest and Range Experiment Station, Ogden, UT.

Burgan, R.E., Shasby, M.B. 1984. Mapping broad-area fire potential from digital fuel, terrain, and weather data. *Journal of Forestry* 82:228–231.

Burrough, P.A. 1986. Principles of geographic information systems for land resources assessment. Clarendon Press, Oxford, England.

Burrows, D.A. 1987. The REFIRES (Regional Fire Regime Simulation) model: A C program for regional fire regime simulation. Unpublished M.A. thesis, University of California, Santa Barbara.

Byrne, R., Michaelsen, J., Soutar, A. 1977. Fossil charcoal as a measure of wildfire frequency in southern California: a preliminary analysis, pp. 361–367. *In* H.A. Mooney and C.E. Conrad (eds.), *Proceedings of the Symposium on the consequences of fire and fuel management in Mediterranean Ecosystems, Palo Alto, California. US Forest Service General Technical Report WO–3.* Washington, DC.

Chuvieco, A.E., Congalton, R.G. 1988. Mapping and inventory of forest fires from digital processing of TM data. *Geocarto International* 4:41–53.

Chuvieco, E., Congalton, R.G. 1989. Application of remote sensing and geographic information systems to forest hazard mapping. *Remote Sensing of Environment* 29:147–159.

Cosentino, M.J., Woodcock, C.E., Franklin, J. 1981. Scene analysis for wildland fire, pp. 635–646. *In Proceedings of the Fifteenth International Symposium on Remote Sensing of Environment.* Environmental Research Institute of Michigan, Anne Arbor, MI.

Davis, F.W., Hickson, D.E., Odion, D.C. 1988. Composition of maritime chaparral related to fire history and soil, Burton Mesa, Santa Barbara County, California. *Madroño* 35:169–195.

Davis, F.W., Borchert, M.I., Odion, D.C. 1989. Establishment of microscale pattern in maritime chaparral after fire. *Vegetatio* 84:53–67.

Deeming, J.E., Burgan, R.E., Cohen, J.D. 1977. The National Fire Danger Rating System – 1978. *USDA Forest Service General Technical Report INT-82*. Intermountain Forest and Range Experiment Station, Ogden, UT.

Frandsen, W.H., Andrews, P.L. 1979. Fire behavior in non-uniform fuels. *USDA Forest Service Research Paper INT-232*. Inter-Mountain Forest and Range Experiment Station, Ogden, UT.

Gibson, L.D., Lenzmeier, C. 1981. A hierarchical pattern extraction system for hexagonally sampled images. *US Air Force Office of Scientific Research Report No. AFOSR-TR-81-0845*. Interactive Systems Corporation.

Gibson, L.D., Lucas, D. 1982. Vectorization of raster images using hierarchical methods. *Computer Graphics and Image Processing* 20:82–89.

Green, D.G. 1983. Shapes of simulated fires in discrete fuels. *Ecological Modelling* 20:21–32.

Green, D.G. 1989. Simulated effects of fire, dispersal, and spatial pattern on competition within forest mosaics. *Vegetatio* 82:139–153.

Green, D.G., Gill, A.M., Noble, I.R. Fire shapes and the adequacy of fire-spread models. *Ecological Modelling* 20:33–45.

Hanes, T.L. 1971. Succession after fire in the chaparral of southern California. *Ecological Monographs* 41:27–51.

Heinselman, M.L. 1973. Fire in the virgin forests of the Boundary Waters Canoe Area, Minnesota. *Quaternary Research* 3:329–382.

Horton, J.S., Kraebel, C.J. 1955. Development of vegetation after fire in the chamise chaparral of southern California. *Ecology* 36:244–262.

Jakubauskis, M.E., Lulla, K.P., Mausel, P.W. 1990. Assessment of vegetation change in a fire-altered forest landscape. *Photogrammetric Engineering and Remote Sensing* 56:371–377.

Johnson, E.A. 1979. Fire recurrence in the subarctic and its implications for vegetation composition. *Canadian Journal of Botany* 57:1374–1379.

Johnson, E.A., Fryer, G.I. 1987. Historical vegetation change in the Kananaskis Valley, Canadian Rockies. *Canadian Journal of Botany* 65:853–858.

Johnson, E.A., Van Wagner, C.E. 1985. The theory and use of two fire history models. *Canadian Journal of Forest Research* 15:214–220.

Keane, R.E., Arno, S.K., Brown, J.K. 1990. Simulating cumulative effects in Ponderosa pine/Douglas fir forests. *Ecology* 71:189–203.

Kilgore, B.M. 1981. Fire in ecosystem distribution and structure, pp. 58–59. *In* H.A. Mooney et al. (eds.), *Proceedings of the Conference on Fire Regimes and Ecosystem Properties. US Forest Service General Technical Report WO-26*. Washington, DC.

Knight, D.H. 1987. Parasites, lightning, and the vegetation mosaic in wilderness landscapes, pp. 59–83. *In* M.G. Turner (ed.), *Landscape heterogeneity and disturbance*. Springer-Verlag, New York.

Lowell, K.E., Astroth, J.H. 1989. Vegetative succession and controlled fire in a glades ecosystem. *International Journal of Geographic Information Systems* 3:69–81.

Lu, J., Bomba, P., Kind, T. 1990. A microcomputer-based geographic information system for forest fire management, pp. 180–192. *Technical Papers of the 50th Annual Convention of the American Society for Photogrammetry and Remote Sensing, Denver, Colorado*.

Malanson, G.P. 1984. Fire history and patterns of California coastal sage scrub. *Vegetatio* 57:121–128.

Minnich, R.A. 1983. Fire mosaics in southern California and northern Baja California. *Science* 219:1287–1294.

Minnich, R.A. 1987. Fire behavior in southern California chaparral before fire control: The Mount Wilson burns at the turn of the century. *Annals of the Association of American Geographers* 77:599–618.

Naveh, Z., Whittaker, R.H. 1979. Structural and floristic diversity of shrublands and woodlands in northern Israel and other Mediterranean areas. *Vegetatio* 41:171–190.

Ogden, G.L. 1975. Differential responses of two oak species to far inland advection of sea-salt spray aerosol. Unpublished Ph.D. dissertation, University of California, Santa Barbara.

Parsons, D.J. 1976. The role of fire in natural communities: an example from the southern Sierra Nevada, California. *Environmental Conservation* 3:91–99.

Rothermel, R.C. 1972. A mathematical model for predicting fire spread in wildland fuels. *USDA Forest Service Research Paper INT–115*. Inter-Mountain Forest and Range Experiment Station, Ogden, UT.

Rothermel, R.C., Philpot, C.W. 1973. Predicting changes in chaparral flammability. *Journal of Forestry* 71:640–643.

Rundel, P.W., Parsens, D.J. 1979. Structural changes in chamise (*Adenostema fasciculation*) along a fire-induced age gradient. *Journal of Range Management* 32:462–486.

Sadar, S.A., Linden, D.S., McGuire, M. 1982. Fuels mapping from Landsat imagery and digital terrain data and fire suppression decisions, pp. 345–351. *Proceedings of the American Congress of Survey and Mapping, Fort Lauderdale, Florida.*

Shasby, M.B., Burgan, R.E., Johnson, G.R. 1981. Broad area Forest and topography mapping using digital Landsat and terrain data, pp. 529–537. *Proceedings of the Seventh Symposium on Machine Processing of Remotely Sensed Data, Purdue University.*

Sweeney, J.R. 1956. Responses of vegetation to fire. *University of California Publications in Botany* 28:143–216.

Trabaud, L., Lepart, J. 1980. Diversity and stability in garrigue ecosystems after fire. *Vegetatio* 43:49–57.

Turner, M.G. 1989. Landscape ecology: The effect of pattern on process. *Annual Review of Ecology and Systematics* 20:171–197.

Turner, M.G., Gardiner, R.H., Dale, V.H., O'Neill, R.V. 1989. Predicting the spread of disturbance across heterogeneous landscapes. *Oikos* 55:121–129.

Turner, M.G., Romme, W.H. 1990. Landscape dynamics in crown fire ecosystems. *In* R.D. Raven and P.N. Omi (eds.), *Pattern and Process in Crown Fire Ecosystems.* Princeton University Press. Princeton, NJ.

Van Wagner, C.E. 1978. Age-class distribution and the forest fire cycle. *Canadian Journal of Forest Research* 7:23–34.

Van Wagtendtonk, J.W. 1985. Fire suppression effects on fuels and succession in short-fire-interval wilderness ecosystems, pp. 119–126. *In* J.E. Lotan, B.M. Kilgore, W.C. Fischer, and R.W. Mutch (Technical Coordinators), *Proceedings of the Symposium and Workshop on Wilderness Fire. USDA Forest Service General Technical Report INT–182.* Inter-Mountain Forest and Range Experiment Station, Ogden, UT.

Yool, S.R., Eckhardt, D.W., Estes, J.E., Cosentino, M.J. 1985. Describing the brushfire hazard in southern California. *Annals of the Association of American Geographers* 75:417–430.

8. Perspectives on Fire Management in Mediterranean Ecosystems of Southern California

Philip J. Riggan, Scott E. Franklin, James A. Brass, and Fred E. Brooks

San Dimas Canyon seems a wild place beyond the reach of civilization. It is home to black bears, gray foxes, Anna's hummingbirds, scrub jays, and in early summer, a multitude of biting insects. Along the steep, north-facing hillsides, the chaparral has the appearance of an ancient forest. From within the canyon it is difficult to remember that one is less than 7 km from metropolitan Los Angeles. It is also difficult to conceive of the landscape swept by flames 30- or 40-m high, or to visualize San Dimas Creek afterwards scoured by debris flows. Our difficulty in perceiving these catastrophic events makes it difficult to alter their course and consequences because to do so involves substantial cost and risks.

Catastrophic fires are undoubtedly part of the natural environment in southern California, given its Mediterranean-type climate and the flammable chaparral that results. This natural flammability, compounded by the common coincidence of human-caused ignitions with high winds and temperatures, has in recent decades produced a series of large and exceptionally severe fires. Since 1960, wildfires have consumed more than 350,000 ha in Los Angeles County (Los Angeles County Fire Department, records on file). There the larger fires dominate the landscape. Since 1919, the largest 10% – those greater than 2400 ha – have accounted for five-sevenths of the total area burned (Figure 8–1).

Fire prevention and suppression have undoubtedly reduced losses to a fraction of those that society and the environment would otherwise have sustained. But the immediate losses from recent wildfires have been so

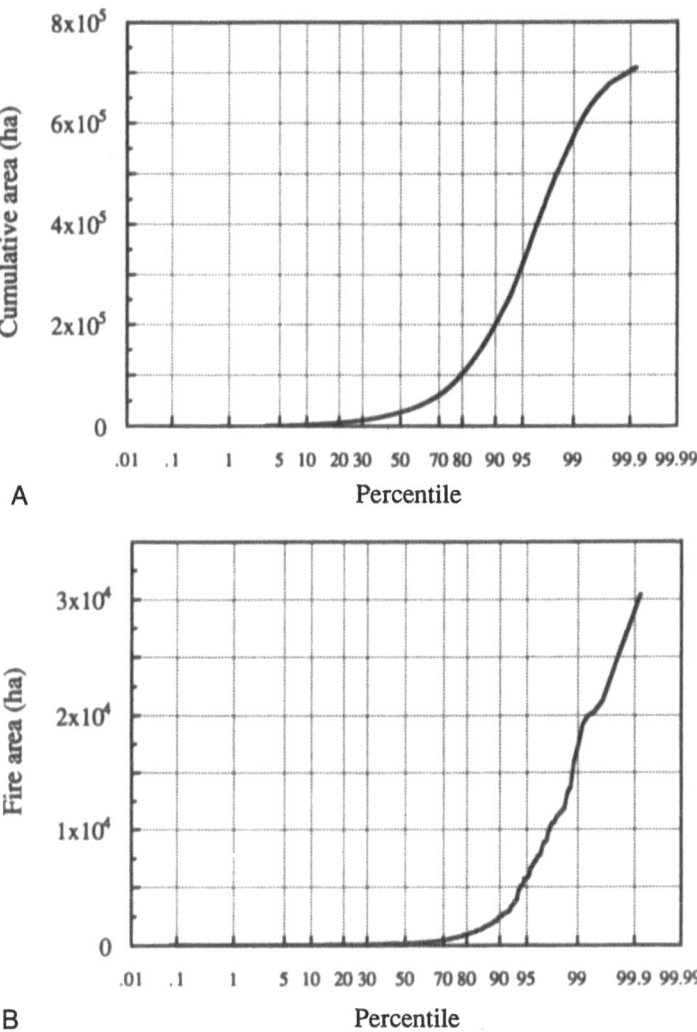

Figure 8–1. (**A**), Cumulative area in Los Angeles County, CA, burned by all wildfires of area greater than 40 ha, 1919 to 1990, expressed as a percentile of fire number from smallest to largest (Los Angeles County Fire Department, unpublished records on file). (**B**), Distribution of individual fire sizes as in *A*. These data show that larger fires have been dominant. Over the period, the largest 10 percent – those greater than 2400 ha – have accounted for five-sevenths of the total area burned.

high as to show that the current balance between ignitions and suppression is unacceptable. One needs only to cite the case of the Paint Fire, in Santa Barbara County, which was ignited late in the afternoon of June 26, 1990. It consumed 1800 ha of chaparral and planted landscaping and destroyed or heavily damaged 648 residences and unattached structures, 221 apartment units, and 11 County buildings (Federal Emergency Management Agency 1990). Most of the destruction occurred during the first evening as local, adiabatically-heated Sundowner winds drove the fire. Estimated costs of the Paint Fire were $250 million in assessed valuation and $2 million for fire suppression (Federal Emergency Management Agency 1990). With continuing urban expansion into wild lands, the losses from catastrophic fires such as this will only accelerate.

There are other serious but less immediate impacts from catastrophic fires. Postfire floods continue to threaten life and property. Fire is a predominant force behind the $750,000 \, m^3$ of soil and rock eroded annually on average in the watershed of the San Gabriel River (S. Kumar, Los Angeles County Department of Public Works, personal communication). This material must be trapped and dealt with before it reaches the urbanized flood plain. Fires accelerate the nitrate pollution of local waters in mountain regions that are subject to heavy air pollution (Riggan et al. 1994a). Frequent burning exacts a toll on native woodlands and has caused some coastal sage scrub communities (for instance, along the mouth of San Gabriel and San Dimas canyons in the San Gabriel Mountains) to yield to invasive European grasses.

Prescribed fire provides a way to mitigate the impacts of catastrophic wildfires by managing the mosaic of natural vegetation. This strategy has been adopted by Federal, state, and local agencies, yet its application is threatened by the primacy of short-term interests over long-term needs.

Ironies abound: In an era of budget cuts, fire suppression is essential, of course, but actions that might reduce the potential for severe fires are not. Fire suppression can be made more efficient by dispatching firefighters nationally, which reduces their availability for prescribed burning during critical months in southern California. Although wildfires emit tremendous quantities of air pollution, they are not conscious acts of society. Prescribed fires are and thus can be regulated and taxed to a point of ineffectiveness. Major wildfires in summer can destroy a multitude of small animals including endangered species, but inadvertent mortality of a smaller number during prescribed burning could be interpreted as a violation of migratory bird treaties and of the Endangered Species Act. When vegetation becomes most flammable from drought and dieback, and prescribed burning is most urgently needed, prescribed burning entails the greatest risk and is least likely to be used. The cumulative impact of legitimate questions about the use of fire could be to so restrict its application as to guarantee the reign of severe fire. Clearly, we need a broader perspective on how we deal – or fail to deal – with fire and ecosystem process in southern California.

Our purpose in this chapter is to examine some of the consequences of managing fire with fire in southern California. These depend on the structure of biomass and the rate it accumulates, the fire behavior that results from that fuel, the relative environmental impacts of moderate-intensity or severe fires, and the potential for intervention.

We argue as follows:

1. Chaparral structure and flammability change year by year during stand development with as much as a threefold increase in energy release during burning between ages 7 and 22 years in some *Ceanothus* communities. Flammability in the *Ceanothus* has also been altered, perhaps dramatically, by climate extremes during the past decade.
2. Postfire flood peaks, sedimentation, and water quality are related to fire severity and should be manageable through the use of prescribed fire to create a mosaic of vegetation that differs in flammability, thereby reducing the likelihood, potential size, or severity of catastrophic fires. Yet, since we cannot readily predict when fire can spread through young chaparral, we cannot specify the level of intervention necessary to achieve substantial reduction in wildfire losses and costs.
3. The concentrations of several radiation-absorbing gases have been rising relentlessly in the atmosphere and raising expectations that global climate will be altered. If climate does change, one of the most important responses to track in southern California will be the potential for catastrophic fires, but this is most affected by extreme conditions that will be difficult to predict.

How Do Chaparral Fuels Develop?

Chaparral fires are fueled by foliage, live stems of small diameter, and deadwood; the spatial arrangement and biomass of these change greatly through the first few decades of postfire community development. *Ceanothus*-dominated chaparral, the development of which has been extensively studied, accumulates above-ground biomass at an exponential rate through at least 2 decades after stand establishment (Riggan et al. 1988). Stand basal area rises exponentially through at least 3 decades (USDA Forest Service records, Riverside, CA). The biomass accumulation depends markedly on site productivity. Biomass of closed-canopy chaparral can range at least from 2 to 12 kg/m^2 at 20 years of age (Figure 8–2). Biomass accumulation rates at older ages are not well established because estimates have been made from only a few chronosequences that confound site factors with age (Marion and Black 1988; Specht 1969).

Foliage biomass of closed-canopy stands of *Ceanothus crassifolius* (hoaryleaf ceanothus) approaches a maximum after 15 to 20 years, although it is likely that there is considerable annual variation about this trend. For instance, we noted that two age classes of foliage were retained through the dry season in 1983, which followed several years of plentiful rainfall (Riggan et al. 1988), yet only one age class was retained through

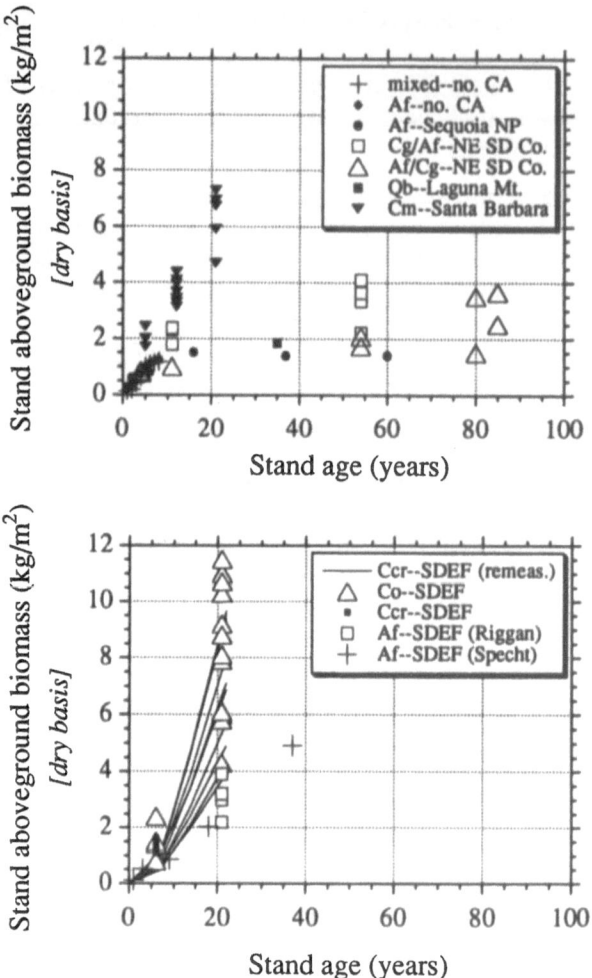

Figure 8–2. Trends in accumulation of above-ground biomass in chaparral. Connected symbols are from reconstructed or remeasured growth. Unconnected symbols represent measurements from stands of different ages within a given region. Production varies markedly across sites, so age does not account for a high proportion of the variance. Stands older than 20 years of age, which are largely from northeastern San Diego County, are not well represented in the data. References are as follows: Cm-Santa Barbara, *Ceanothus megacarpus* (Schlesinger and Gill 1980); mixed and Af-no. CA, mixed chaparral and *Adenostoma fasciculatum* from Mendocino and Shasta counties, CA (Sampson 1944); Af-SDEF, *A. fasciculatum* at San Dimas Experimental Forest (prior to the Johnstone Fire in 1960) (Specht 1969); Ccr, Co, and Af-SDEF, *C. crassifolius*, *C. oliganthus*, and *A. fasciculatum* at San Dimas Experimental Forest (subsequent to the Johnstone Fire in 1960); Qb-Laguna Mt., *Quercus berberidifolia* in southern San Diego County (Riggan et al. 1988); Cg/Af and Af/Cg-NE SD Co. (with species order denoting dominance), *C. greggii*, *A. fasciculatum* (with elements of *A. sparsifolium* (redshanks) and *Quercus berberidifolia*) in northeast San Diego County (Marion and Black 1988); Af-Sequoia NP, *A. fasciculatum* at Sequoia National Park (Rundel and Parsons 1979).

the years of low rainfall from 1984 through 1991. Foliage biomass probably approaches a maximum when the developing leaf area of dominant shrubs hastens the onset of seasonal water stress and the effects of this extended physiological drought are no longer balanced by enhanced carbon uptake during the early growing season or more efficient competition for water with smaller plants (Riggan et al. 1988). Foliage biomass of mature chaparral may also vary substantially across sites; measured values in 21-year-old *Ceanothus crassifolius* stands ranged almost threefold from 0.5 to 1.3 kg/m^2 (Riggan et al. 1988).

As chaparral ages, deadwood slowly accumulates through progressive shading of lower branches, drought- or pathogen-induced declines or dieback, intraspecific competition, and random mortality such as that from snow or freezing. Accumulation rates vary according to the life histories of the predominant species (nomenclature follows that of the Jepson Manual [Hickman 1993]):

Quercus berberidifolia (scrub oak) accumulates little standing deadwood, which constitutes only one-tenth of the above-ground biomass in mature stands 21 to 37 years of age (Riggan et al. 1988).

Adenostoma fasciculatum (chamise) apparently accumulates little deadwood in low productivity environments (0.07 of the total) and a substantial fraction (0.45 to 0.48) in high-productivity stands, especially where subject to competition from tall *Ceanothus*, such as along the coastal slope of the San Gabriel Mountains (Riggan et al. 1988). Median fraction deadwood across 15 southern California locations has been reported to be 0.23 (Paysen and Cohen 1990). Shoot mortality is probably high when a sequence of years with relatively high rainfall are followed by prolonged drought (T. Paysen, personal communication).

Ceanothus crassifolius and *C. megacarpus* (big-pod ceanothus) are little affected by shading and competition. Deadwood in the lower canopy of live *C. crassifolius* at the San Dimas Experimental Forest constitutes less than 0.01 of above-ground biomass at age 6, and 0.12 at age 21. Mortality of small shrubs, which occurs primarily between the ages of 5 to 10 years, can raise the fraction of deadwood to 0.19 of the community's above-ground biomass by age 21 (Riggan et al. 1988). Accumulated deadwood in *C. megacarpus* constitutes 0.23 of standing biomass at this age (Schlesinger and Gill 1980). Several of the Ceanothus species are also subject to high, pathogen-induced mortality rates as we discuss later.

Salvia mellifera (black sage) is a semiwoody shrub of low stature which, in competition with tall *C. crassifolius*, accumulates a substantial biomass of deadwood, constituting as much as one-third of above-ground biomass at age 21 years (Riggan et al. 1988). *Salvia mellifera* can be widely established when postfire regeneration of competing taller shrubs such as *Ceanothus* is poor, thereby increasing the probability of burning at young ages or under moderate weather (Riggan et al. 1988).

Snow or frost is occasionally an important cause of mortality. The 40,000-ha Marble Cone Fire of 1977 was an especially severe chaparral fire fueled by stems broken by a wet, heavy snowfall (T. Plumb, personal communication). Frost also occasionally kills stems and foliage of *Malosma laurina* (laurel sumac), thereby producing scattered dead shrubs as was widely observed in the winter of 1990–91 (D. Neff, California Department of Forestry and Fire Protection, personal communication). This mortality is most likely to affect fire behavior during the limited time that dead foliage is retained on the plant. Exceptionally low temperatures in the absence of an insulating snow cover in the winter of 1990–91 also caused widespread mortality of *Arctostaphylos* spp. in stands regenerating after the 1987 wildfires in northeastern California (California Department of Forestry and Fire Protection, personal communication).

Chaparral Dieback in Southern California

Extensive chaparral dieback may have caused an abrupt increase in fire potential in southern California beginning in 1984 and 1985 (Figure 8–3). This mortality appears to have been most pronounced on coastal slopes in the San Gabriel and the Santa Ynez mountains, in the Santa Monica Mountains, and in southern San Diego County. Accelerated mortality has been most notable in *Ceanothus crassifolius* and *C. megacarpus*, although several chaparral species have been affected.

Climatic Stress

Severe drought is most likely the primary stress factor that predisposed the affected chaparral to infection by a fungal pathogen. We first observed dieback to be extensive in the San Gabriel Mountains after the hydrologic year beginning October 1, 1983, during which precipitation was in the eleventh percentile of a 107-year record (as measured at Tanbark Flats, San Dimas Experimental Forest [USDA Forest Service, Riverside, CA; unpublished records on file]). More important, nine-tenths of the autumn and winter rains that year fell before the end of December; the largest storm during the succeeding 11 months occurred during August and brought only 1.3 cm of rain. The 1983–84 drought followed a hydrologic year with twice the average annual rainfall, during which an abundance of foliage developed. With an extensive leaf area, shrubs probably exhausted available soil water and suffered substantial water stress even during the growing season. The dieback progressed annually through 1992 with a second peak in activity late in 1986 after the fourth lowest annual rainfall on record.

A similar pattern of abundant rainfall, followed by a tenth-percentile drought occurred during hydrologic years 1979–80 and 1980–81; however, no extensive dieback occurred. This lack of response may have been because rain during the low-precipitation year of 1980–81 fell largely from late January through early April, coincident with the growing season.

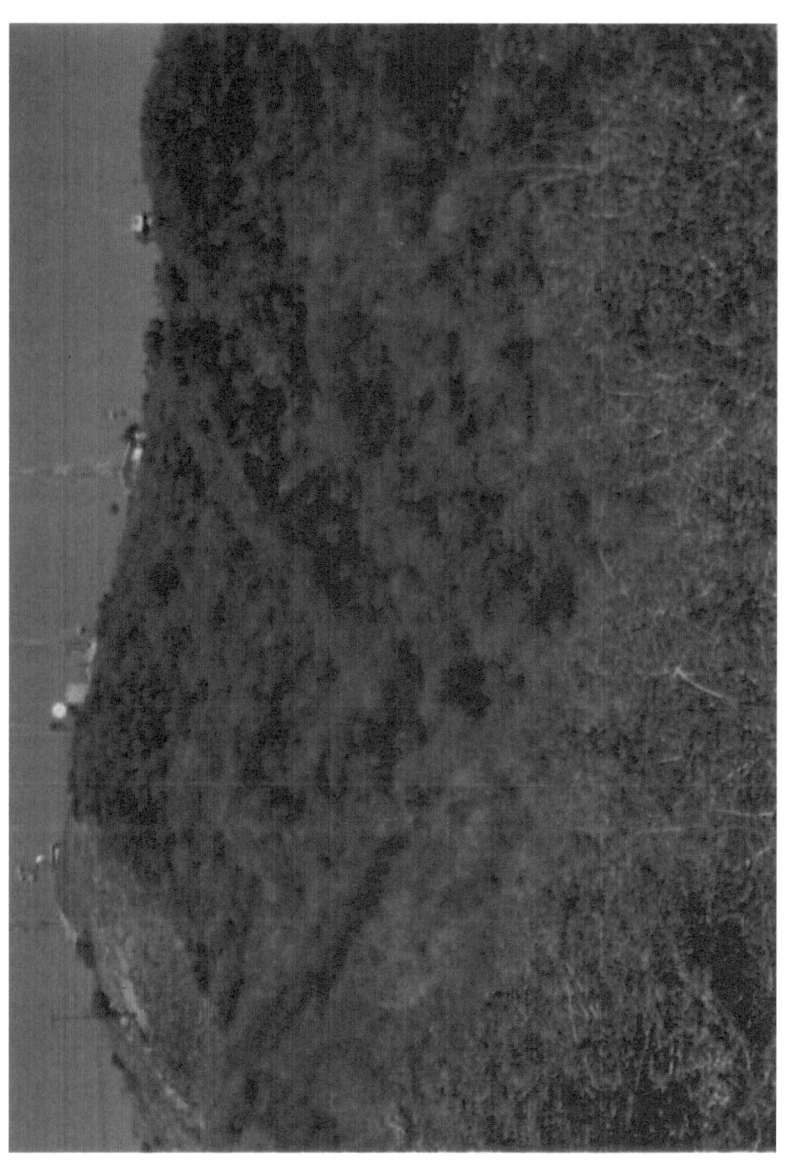

Figure 8–3. Dieback in *Ceanothus crassifolius* in Ham Canyon in the San Dimas Experimental Forest, March 1985.

The initial pattern of dieback may also have been linked with topographic gradients in water stress. We observed in Bell Canyon on the San Dimas Experimental Forest that the dieback was most common on the steepest and driest slopes where *Ceanothus crassifolius* is mixed with *Adenostoma fasciculatum*. Ridge lines dominated by the *Ceanothus* were least affected.

Another unusual weather pattern during August 1984 may have affected the dieback. A subtropical air mass over southern California caused extended high temperatures and humidity with occasional light rain. This rain could have facilitated dispersal of fungal spores as the high temperatures and moisture promoted subsequent secondary infections.

Botryosphaeria dothidea *and Its Mode of Action*

An opportunistic pathogen, *Botryospheaeria dothidea* (or *B. ribis*), was identified in association with dieback of stems of *Ceanothus crassifolius*, *C. oliganthus*, *C. megacarpus*, *C. spinosus*, *C. leucodermis*, *Heteromeles arbutifolia*, *Arctostaphylos glauca*, *A. glandulosa*, and *Quercus agrifolia* (J. Pronos, USDA Forest Service, San Francisco, CA; personal communication). Subsequent inoculation trials have established that *Botryosphaeria dothidea* can reproduce the dieback symptoms (Brooks and Ferrin 1991).

Botryosphaeria dothidea is a common canker fungus (Hodges 1983) that is probably ubiquitous in the chaparral environment. It can attack a variety of hosts, including *Cornus stolonifera* (red osier dogwood) (Schoeneweiss 1979), *Sequoiadendron giganteum* (giant sequoia), *Sequoia sempervirens* (coast redwood) (Worrall et al. 1986), and *Pinus taeda* (Loblolly pine) (Hodges 1983). Chaparral dieback in 1984 and 1985 was coincident with widespread mortality in urban vegetation, including plantings of *Ceanothus* and *Arctostaphylos* at the Rancho Santa Ana Botanic Garden in Claremont, CA. Dieback in chaparral also was apparently widespread circa 1948 to 1951 (Kittredge 1955) and 1958 to 1961 (Pirsko and Green 1967) during periods of record-low rainfall.

Botryosphaeria dothidea is considered a facultative parasite and occurs naturally as a saprophyte on dead and dying plant tissue. Infection proceeds through wounded plant tissue (Schreiber 1964; Worrall et al. 1986) or natural openings such as lenticels (Shearer et al. 1987) or leaf scars (Spiers 1977). Tissue entry does not assure spread of the pathogen. The host is generally attacked when subject to a predisposing stress, which can occur after initial entry. Possible stresses to the host in this instance could be abnormally low water potential (Crist and Schoeneweiss 1975; Schoeneweiss 1979, 1981), prolonged duration of water stress, defoliation (Schoeneweiss 1981), or possibly impact of air pollution (Hinrichsen 1987; Schoeneweiss 1975). Plants do mount defenses to the spread of the pathogen within a stem, as shown by the canker, but this defense may be inadequate under prolonged stress conditions. Fungal hyphae progressing through vessel elements in a stem restrict water movement to the distal leaves. Once the hyphae infect a stem below the point at which live leaves are attached, the entire stem is killed.

Levels of Mortality

We estimated the accumulation of deadwood in *Ceanothus crassifolius* chaparral after the initial dieback at Lodi and Ham canyons at the San Dimas Experimental Forest (Basal area was measured 50 cm from the base of each stem found in 33 plots, each 120 m^2 in area, in which *Ceanothus crassifolius* accounted for more than one-half of the basal area. The fraction of deadwood within live stems was measured for a sample of 183 stems destructively sampled in 1986 and 1988 at Lodi and Ham Canyons; variance of the ratio estimate was 4.8×10^{-4}. Woody tissue at any point was classified as live if any live foliage remained distal to that point along a branch.) Deadwood constituted 0.12 of the woody biomass within live stems. Dead stems accounted for an average of one-quarter of the total basal area. Thus, cumulative mortality affected three-eighths of the woody biomass.

Relation to Stand Age

Chaparral stands older than 30 to 40 years have been described as senescent, a condition marked by mortality of individual shrubs and loss of canopy continuity (Specht 1969; Hanes 1971). For example, Hanes (1971) observed in the San Gabriel and the San Bernardino mountains in 1966 and 1967 that over one-half of the population of *Ceanothus crassifolius* and *C. greggii* var. *vestitus* was dead in stands older than 40 years. Dead shrubs constituted 70% of the total *Ceanothus crassifolius* biomass in 37-year-old mixed-composition stands measured in 1956 at the San Dimas Experimental Forest (Specht 1969). Yet, high rates of mortality are not a universal characteristic of older Ceanothus stands. Schlesinger and Gill (1980) observed 54-year-old *Ceanothus megacarpus* that had no greater mortality than stands less than one-half that age. We surmise that mortality previously attributed to senescence or to "intraspecific competition" may actually be driven by *Botryosphaeria dothidea* or other pathogens. Furthermore, the dieback does not appear to be merely a result of chaparral shrubs exceeding their natural longevity because we have observed it in 6-year-old *C. megacarpus* in the Santa Monica Mountains.

How Is Fire Behavior Affected by Biomass Accumulation and Dieback?

Several components of fire behavior are important to managing fire effects. These include likelihood that fire will spread, total energy release, forward velocity or rate of spread of the fire front, and fire intensity. Rate of spread affects the ultimate size of a fire given concerted effort at suppression and a finite duration of severe weather. Unfortunately, there are no direct, quantitative observations of how fire behavior changes with either stand development or dieback. However, we can show some qualitative effects of stand age from our observations of the July 1985 Sherwood Fire in western Los Angeles County, and deduce some properties from fuel distributions and, by analogy, model fires in porous fuel beds.

Observations of Fire Spread in Two Age Classes

The Sherwood Fire burned concurrently in 50- and 8-year-old chaparral under the influence of a subtropical high aloft. (Afternoon temperatures were 38 to 40°C, relative humidity was below 20%, and winds were from the northwest at approximately 15 km/h.) Before burning, the older age class had had a several-fold greater biomass than the young, as judged from vegetation at the fire perimeter. Efforts at fire suppression were minimal in the young age class until the late stages of the fire.

Fire in the young chaparral produced a mosaic of burned and unburned vegetation, spreading in part through the *Salvia mellifera* and grass fuel embedded in this *Adenostoma fasciculatum* and *Ceanothus megacarpus* community. There fire consumed foliage and only small live twigs, incompletely consumed the leaf litter, and left a black ash surface. Radiometric temperatures of ash under solar heating 3 weeks after the fire were as high as 72°C (Figure 8–4A).

In stark contrast, the older chaparral burned with exceptionally high intensity as all foliage, fine wood, leaf litter, and many live stems 5- to 10-cm in diameter were incinerated. Virtually no unburned vegetation remained within the fire perimeter in this age class, and the ground surface was uniformly covered with a fine white ash. This ash was approximately 10°C cooler under solar heating than that of the young age class, and it was more reflective at intermediate wavelengths of infrared light (Figure 8–4B). We interpret the temperature and reflectance differences as showing greater carbon loss from soils and leaf litter under the older vegetation. This illustrates that both the propensity of chaparral to propagate fire and a fire's severity are related to changes in biomass associated with stand aging.

The vegetation mosaic created in young chaparral by the Sherwood Fire shows, for wildfire conditions, a general observation from prescribed burning: Chaparral fires spread only when the rate of energy transfer to unburned fuel exceeds a relatively substantial threshold. Weather that allowed a severe fire in older chaparral was only marginally sufficient to spread fire in the young chaparral. Such marginal conditions during prescribed burning, as judged by an inability to sustain combustion on shallow or shaded slopes or at any but the lowest live-fuel moisture, have produced large flames of 6 to 10 m in 35-year-old *Quercus berberidifolia* (live fuel moisture, 0.8 g water g^{-1} dry mass; wind run, 4.8 km/h with gusts to 24; relative humidity, 21 to 27%) (Dougherty and Riggan 1982) and 30 m in 20- to 25-year-old *Ceanothus crassifolius* (Live-fuel moisture, 0.7 to 0.8 g water g^{-1} dry mass; wind gusts to 15 km/h) (Riggan et al. 1994a). In the absence of substantial deadwood biomass, fire is readily spread only with wind and low live-fuel moisture.

Total Energy Release During Burning

As we have noted, chaparral fires typically consume a fraction of live wood, most leaf litter, and all foliage and deadwood. For a given stand, the fraction of live wood consumed apparently varies little across the range of weather and fuel moisture that allows fire propagation (Figure 8–5). In 22-year-old *Ceanothus crassifolius*, 0.09 of the live wood was lost, an amount that corresponded to six-tenths of the wood less than 0.5 cm in diameter. If we assume that fires in younger stands consume the

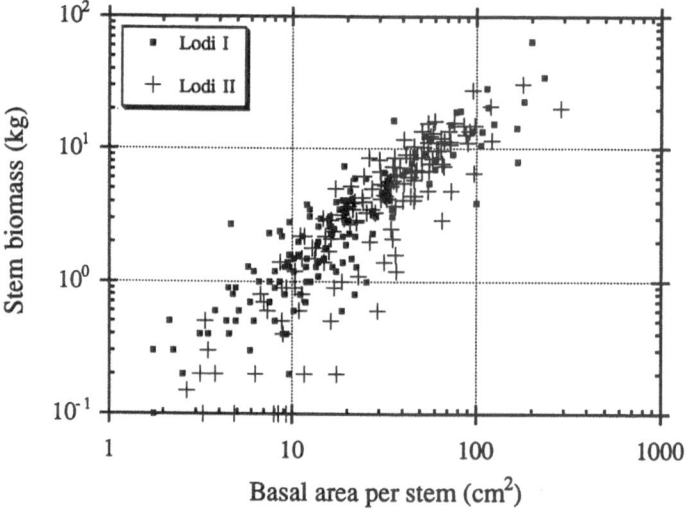

Figure 8–5. Relation of unconsumed biomass to stem basal area after two pre-scribed fires in *Ceanothus crassifolius* chaparral during December 1986 and June 1987 at Lodi Canyon in the San Dimas Experimental Forest. Basal area of live stems was not reduced by burning, and there was no discernible difference between the fires in the rate of biomass consumption per stem. The winter fire (Lodi I) was ignited with difficulty and only spread slowly on steep slopes and southerly aspects; the summer fire (Lodi II) spread readily on similar slopes, yet both fires consumed approximately 0.09 of the live-wood biomass and equivalent masses of foliage and deadwood. Thus, despite large differences in rate of spread and flame length, the fires did not differ in total energy release.

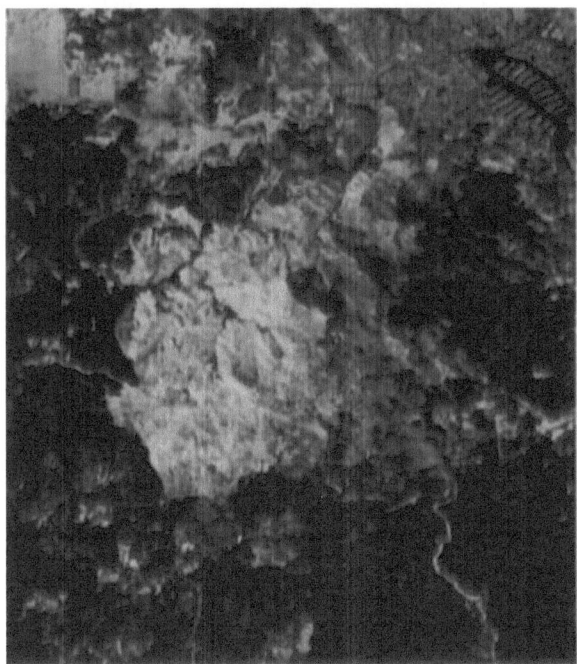

Figure 8–4. *A*, Ash from the 1985 Sherwood Fire in the western Santa Monica Mountains, Los Angeles and Ventura counties, CA. In this false-color composite image, reflected light at 1.55 to 1.75 µm is mapped in blue and that at 2.08 to 2.35 µm is mapped in green; emitted long-wave infrared light at 10.4 to 12.5 µm is mapped in red. Radiometric temperatures of the surface ash, estimated from the 10.4 to 12.5 µm radiance, were as high as 72°C in the younger age class and 10°C cooler in the older age class. The eastern one-third of the fire area (tones in orange at the right) is the burned 8-year-old age class. The yellow tones in the northwest quadrant of the fire area are from a north-facing aspect where 50-year-old chaparral was burned. Slopes at the southwest (below the fuel break that appears as an east-west trending white line) were south-facing and of intermediate age. Data were collected 23 July 1985 by an ADDS 1268 Thematic Mapper Simulator (Daedalus Corporation) aboard a NASA high-altitude aircraft.

(*Continued*)

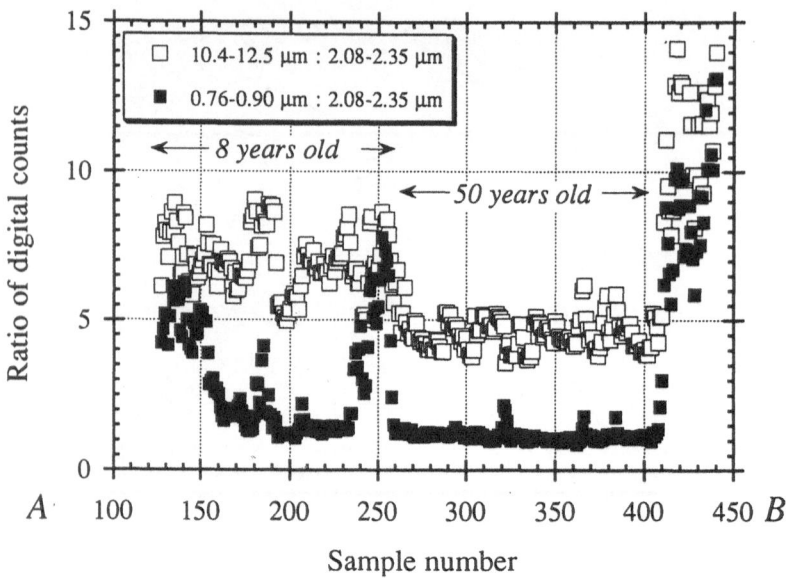

Figure 8–4 (*Continued*). *B*, Graphical representation of the reflected and emitted light along a transect from east to west through the fire area depicted in *A*. Fire in the 8-year-old chaparral (sample numbers 120 to 260) left a mosaic of ash and unconsumed vegetation, which is marked by high values of the ratio of near-infrared (0.76 to 0.90 μm) to intermediate-infrared radiances (2.08 to 2.35 μm). Ash there was warmer and less reflective at intermediate infrared wavelengths (depicted by high values of the ratio of radiances at 10.4 to 12.5 μm and 2.08 to 2.35 μm) than was the ash from the older age class.

same fraction of live wood in this size class, then fire spreading in 7-year-old stands of this species, necessarily burning with low live-fuel moisture under considerable winds, would yield from standing biomass approximately one-tenth of the energy of a fire burning in a healthy stand at age 22 years (Figure 8–6). In each case, the heat of vaporization would be small in relation to the total energy released.

Foliage, soil O horizon (organic material at the soil surface), and consumed wood at age 22 years are roughly equivalent sources of carbon for combustion (Table 8–1). Thus, the O horizon adds energy to the fire equivalent to one-half of that from the standing vegetation, although it will not propagate fire by itself because it is too compact.

Dieback of one-half (a high rate) of the live stems in a 22-year-old stand of *Ceanothus crassifolius* might be expected to increase energy release by one-half compared with the same stand in a healthy condition (see Table 8–1).

Changes in Rate of Spread During Stand Development

It is difficult to translate biomass consumption or energy release from these cases into predicted rates of fire spread because to do so would

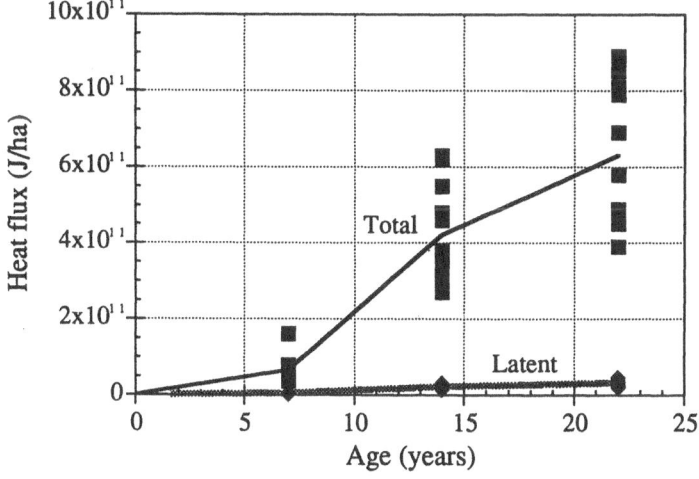

Figure 8–6. Estimated total and latent heat fluxes during burning of standing biomass in 7-, 14-, and 22-year-old *Ceanothus crassifolius*. Estimates were based on reconstructed biomass trends (Riggan et al. 1988) and the following assumptions: live wood consumed (0.09 of the total) is a constant fraction (0.6) of wood with diameter <0.5 cm; by analogy to *Quercus berberidifolia*, wood with diameter <0.5 cm is one-quarter of the total at age 7 and one-fifth of that at age 14; moisture contents are 0.7 g water g^{-1} dry biomass for foliage and live wood and 0.07 g water g^{-1} dry biomass for deadwood; specific heat of vaporization is 2.26×10^6 J/kg; specific heat of combustion is 2×10^7 J/kg (Albini 1980).

Table 8–1. Estimated Biomass Consumption During Burning for 22-Year-old Stands of *Ceanothus crassifolius* Both With and Without An Additional 50% Mortality of Stems

Component	Biomass Consumption (Mg/ha)		Carbon Loss (Mg/ha)	
	No Dieback	50% Dieback	No Dieback	50% Dieback
Live biomass				
Foliage	14.3	7.2	7.2	3.6
Live Wood	4.0	2.6	2.0	1.3
Deadwood				
Within live shrubs	8.0	30.3	4.0	15.2
Dead shrubs	4.2	4.2	2.1	2.1
Readily consumed (<0.5 cm)	4.6	7.9	2.3	4.0
O horizon	45.5	52.7	6.4	10.0
Total consumed			21.7	32.2
Readily consumed			17.9	18.9

Note. Biomass is reported by Riggan et al. 1988. Carbon loss from the O horizon was estimated from measured carbon fraction there (mean = 0.24, s = 0.081, n = 72) and in ash (mean = 0.096, s = 0.038, n = 61) after prescribed fires in standing chaparral at San Dimas in 1984; the case with dieback assumes consumption of abscised foliage before appreciable decomposition could begin. Readily consumed deadwood biomass is assumed to be that with a diameter less than 0.5 cm; larger material may also be consumed within the flaming front.

require a questionable extrapolation from semiempirical descriptions of model fires in a laboratory. Some inferences may be possible, however, from two models: One described by Rothermel (1972) and elaborated by Wilson (1990), and one of Fendell et al. (1990).

Rothermel (1972) gives a semiempirical model of fire spread that was parameterized first for small fires burning with no wind in dry excelsior or in wood elements with a nominal width of 1/4 or 1/2 in. The fuel array was approximately 90-cm wide and 11- to 15-cm deep. Rate of spread with wind was then described by $V(U) = f(V_0 (1 + f_w))$ where V is the forward rate of spread at wind speed U, V_0 is the spread rate with no wind, and f_w is a function of wind speed, relative surface area to volume of the fuel, and the fuel packing, b, which can be interpreted as the proportion of the fuel bed volume that is actually occupied by fuel elements. The wind speed function, f_w, was estimated from observations of grass fires in Australia, in which no fuel properties were measured, and from limited trials with fine fuels in the laboratory.

This model was not explicitly developed to describe fire spread in sparse live fuels such as those that propagate chaparral fires. The form of the model (which relates spread with wind to a finite rate with no wind) is also inappropriate for chaparral, where there is commonly a threshold of wind required for fire propagation (i.e., the no-wind rate of spread is zero).

Fendell et al. (1990) solved a differential equation for fire spread driven by convective heating. The resulting model, $V(U) = a(U/m)^{1/2}$ (where a is an experimentally determined constant), is simple in form and explicitly includes interaction of wind and mass, m, for fuel of a given height and ratio of surface area to volume. The model parameter, a, was estimated from burning trials conducted with ranges of wind speed (up to 4.6 m/s) and mass loading of toothpicks. The ratio of surface area to volume for toothpicks is similar to that of chaparral foliage, but the packing used experimentally was much denser than that of chaparral. The experiments used combustion tables of 55- or 100-cm width and incorporated variations in fuel height below 22 cm. For most experiments, which were conducted with a constant height of fuel, the changes in mass were equivalent to a change in packing.

These models are limited by the experimental situations from which they were developed. Spread rate of the model fires was dependent on width of the fuel bed and the magnitude of edge effects on forward energy transfer, roughly doubling as the width was increased from 55 to 100 cm (Fendell et al. 1990). Fendell et al. (1990) also generally observed reduced rates of spread when b and U are held constant, but the mass is increased by using taller fuel elements. This reduction presumably resulted from dispersal of convective energy through a larger volume of air within the fuel bed. Thus, velocity of fire spread was dependent on the geometry of the experimental situation, and there is no way to know a priori how to scale the results from model fires to those of chaparral fires where flame lengths are typically one to two orders of magnitude greater; large live woody stems do not burn but reduce energy transfer to unburned fuel; and there is substantial oxygen depletion, fire-caused acceleration of ground-level winds, and turbulence.

Both sets of experimental results show that rate of spread in fine fuels reaches a maximum at low values of packing. Furthermore, the experimental results of Fendell et al. (1990) show that the packing that maximizes spread obtains lower values at higher wind speeds. Given appreciable wind, fire spread was very sensitive to the packing at levels below the optimum (which can act as a threshold for spread), and zero spread was observed at finite values of fuel loading. (The model describes only conditions with packing greater than the optimum.)

We assume by analogy to the laboratory-based fires that chaparral with low deadwood biomass, which may only marginally sustain fire spread at lower wind speeds, is essentially at an optimum packing for those conditions. Furthermore, we assume that an increase in foliage mass as a stand matures (e.g., between age 10 and 20 years) increases packing above this optimum because of greater overlap between the canopy of adjacent shrubs and slow concurrent height growth. Therefore, the rate of spread would be expected to decline accordingly. Furthermore, if the theoretical model form proposed by Fendell et al. (1990), $V = f(U/m)^{1/2}$, remains valid at the low packing and requisite wind speeds of chaparral fires, then the accumulation of foliage, fine live wood, and fine deadwood (as given in Table 8–1 and Figure 8–6) between ages 14 and 22 years would reduce the forward rate of spread by about one-fifth.

The impact of the dieback on rate of spread is more problematical because we cannot readily predict, for the combined live and dead fuels, the interactions of wind, packing, and moisture or the effect of fuel stratification. In general, we expect that the primary effects will be to increase fire residence time and soil heating, because of the greater total energy release and larger-diameter fuels that become involved, and to allow fire spread at lower wind speeds and higher fuel moistures. The latter effect could extend, perhaps by several weeks in early summer or late autumn, the conditions under which large fires may occur or could increase fire size during more severe weather – for example, by accelerating fire spread at night.

How Are Fire Impacts Related to Fire Severity?

Flooding and Sedimentation

Destructive postfire flooding, such as that which caused 30 deaths and destroyed 484 homes in the towns of La Cresenta and Montrose in 1933, has historically been an impetus for fire and flood control in southern California (Eaton 1935). An extensive system of engineered debris catchments and flood channels has been constructed to contain such flows, but these provide limited protection even today (S. Kumar, Los Angeles County Department of Public Works, personal communication). Debris flows from recently burned chaparral watersheds can mobilize a large volume of material from sand to boulders of 1- to 2-m in breadth. When the capacity of a debris basin has been exceeded, this high-velocity mix cuts its own unpredictable and destructive path. Even when the flows are contained, the expense of emergency sediment removal is high.

The severity of a chaparral fire can affect the magnitude of subsequent flooding and sediment production. This was demonstrated in 1984 and 1985 in a replicated watershed experiment at the San Dimas Experimental Forest (Riggan et al. 1994a, 1994b), where two canyons burned by moderate-intensity prescribed fires in standing chaparral yielded three-eighths as much sediment and water as did two that had been burned by severe fires in felled *Ceanothus*. The severe fires consumed all of the above-ground biomass, thereby releasing approximately three times the energy of the more moderate fires. The postfire sedimentation was dominated by debris flows that were generated at a rainfall intensity of 5 mm in 5 minutes and comprised 59% sediment by weight (Weirich 1988; Riggan et al. 1994b). Associated peak discharge from the severely burned canyons exceeded $330 \, L \, s^{-1} \, ha^{-1}$; concurrent flows peaked in unburned watersheds at less than $1 \, L \, s^{-1} \, ha^{-1}$.

Massive soil erosion after intense chaparral fires probably follows from the loss of vegetative cover, organic matter destruction, and development of water repellency in soils. Loss of organic matter on and beneath the soil surface probably parallels the intensity and depth of soil heating and contributes to sediment production by exposing mineral soil and reducing its cohesiveness. Water repellency forms in soils when pyrolytic organic compounds condense below the soil surface and block subsequent infiltration of water (DeBano 1981). The depth of this condensation, as well as that of initial rill and surface erosion, is probably greatest under prolonged or intense soil heating (Wells 1981).

Nitrogen Flux from Burned Watersheds

Fire also affects the accumulation and cycling of mineral nitrogen in soils and the flux of nitrate in stream water. In the San Dimas experiments (Riggan et al. 1994a), burning mobilized nitrogen that had accumulated in part from the air pollution of metropolitan Los Angeles and caused stream water to be polluted with nitrate (NO_3^-) at concentrations sometimes exceeding the Federal water quality standard (0.7 meq/L). Stream-water NO_3^- concentrations were elevated during peak flows, reaching 1.12 meq/L during a debris flow.

The postfire NO_3^- concentrations and flux in stream water reflected the severity of burning. Severe fires produced daily-mean volume-weighted NO_3^- concentrations that were 1.7 times those after the moderate-intensity fires; moderate fires produced concentrations that were 3.0 times those of unburned controls. Annual NO_3^- loss from severely burned watersheds, averaging 1.2 keq/ha, was 40 times greater than that from areas that remained unburned. Fires of moderate intensity produced less than one-seventh the NO_3^- loss observed after severe burning.

The response of stream-water NO_3^- to burning was due in part to changes in the net rates of soil mineralization and nitrification, both of which were positively associated with fire severity (Riggan et al. 1994a). Severe fires undoubtedly subjected soils to the greatest heating below the immediate soil surface and caused the greatest rates of nitrogen volatilization. As a result, they elevated the NH_4^+ concentration in surface soils, but to a level below that of the moderate fires. NH_4^+ concentrations were subsequently highest in soils that had been subject to severe heating, because of either long-term inhibition of NH_4^+ immobilization or accelerated mineralization. Nitrate concentration after storms in November and December was greatest in soils subjected to severe fire (1.20 meq/kg of soil), intermediate in those from the moderate-intensity fire (0.79 meq/kg), and least in the unburned soils (0.49 meq/kg).

The results from the San Dimas fire severity experiments show that flooding, sedimentation, and nitrate water pollution after a moderate-intensity prescribed fire are likely to be substantially less than would be produced from an equivalent area burned by a severe, catastrophic wildfire. The experiment was designed to produce a range in severity between fires, and although a fuel comprised entirely of deadwood would rarely be found in nature, the difference in energy release rate between the fires is comparable to that we estimate between fires burning in 10- and 22-year-old *Ceanothus crassifolius*.

Fire Severity and Community Composition

Fire severity can also affect the subsequent species composition of chaparral with potentially long-lived effects. This was apparently the case after prescribed fires we examined at Stone Canyon in the Santa Monica Mountains where *Ceanothus megacarpus* seedling density 1 year after burning was 5 times higher where the chaparral had been crushed and burned than where standing vegetation had been burned. (Seedling density was estimated in 46 plots each $10 \, m^2$ in area; average density was $22.5 \, m^{-2}$ after crushing and burning and $4 \, m^{-2}$ after burning in standing vegetation. The null hypothesis that densities were not different was tested with a Mann–Whitney test and rejected with $P = 0.01$.) As at San Dimas, crushing allowed virtually all of the above-ground biomass to be burned with correspondingly greater energy release.

Species abundance in mature stands at the San Dimas Experimental Forest also shifted dramatically after successive fires of widely differing severity in 1919 and 1960 (Jacks 1984). The cover of *Ceanothus crassifolius* in 15-year-old stands after the 1960 Johnstone Fire was six times that which had been estimated at a comparable age in the same area after the 1919 San Gabriel Fire (Horton 1941), whereas cover of *Adenostoma fasciculatum* was one-third as great and the amount of bare ground was reduced. Changes in cover were primarily due to an increase in the Ceanothus population density from 0.03 to $0.14 \, m^{-2}$; the mature *Adenostoma fasciculatum* population concurrently declined from 0.06 to $0.04 \, m^{-2}$. The Johnstone Fire must have had relatively high energy yield because it burned 41-year-old chaparral that had been subjected to substantial mortality during an interval of record drought (Kittredge 1955); the San Gabriel Fire burned in 24-year-old stands. Drought after the Johnstone Fire may also have enhanced *Ceanothus crassifolius* seedling survival relative to that of *Adenostoma fasciculatum* (Jacks 1984).

Fuel structure and biomass accumulation depend to a large extent on species composition, so a change in composition could alter the nature of the next fire. An increase in *Ceanothus* abundance, such as that which occurred at San Dimas after 1960, could yield a structure more conducive to the propagation of infrequent, severe fires (Riggan et al. 1988).

How Might Prescribed Burning Alter the Pattern of Catastrophic Fires?

Now, let us consider some of the values and environmental liabilities that might be derived from an active program of prescribed burning. The strategy is to maintain a shifting mosaic of age classes both to alter the potential for catastrophic fire and to reduce overall environmental impacts of an unmanaged wildfire regime. Our analysis is somewhat speculative because the ability of any chaparral community to propagate fire at a given age or fuel structure is probably the least known property of fire behavior.

Our approach is to ask what might result when a wildfire ignites and spreads in mature chaparral, 30 or 40 years of age, and encounters a young age class. Let us assume that it is autumn with low fuel moisture, humidity near 20%, and moderate winds (under 20 km/h).

The Wildfire Front is Large Compared to the Size of the Younger Age Class. In this case, the ultimate *perimeter* and rate of progress of the wildfire will not be materially reduced, but there are several alternative outcomes within the younger chaparral.

If the young chaparral is 2 to 5 years old (and not dominated by introduced grasses), then it is likely that it will not burn and the overall area of the wildfire will be reduced accordingly. Such was the case where the 1985 Wheeler Fire encountered the 1983 Matilija Fire (Figure 8–7). Considering the wildfire and prescribed fire together, the total area burned will not be affected, but there will be a reduction in watershed impacts within the area of the prescribed burn, and a partitioning of impacts between the years after the two fires. Thus, within the uncertainty of ensuing storm intensity, there should be reduced flood peaks and less sediment and debris to deal with in an emergency.

If the young age class is 10 to 15 years old, there is a greater, but still small, chance that it will propagate the wildfire. If it does burn, biomass consumption and total energy release within the young stand would probably be one-quarter to one-half that which would have resulted without management, and the relative reduction in subsequent flood peaks and sedimentation from the area might be roughly comparable to that observed between moderate-intensity and severe fires in the San Dimas experiments. The result of the two fires together would be an increase in fire frequency within the perimeter of the prescribed fire, but a small reduction in sedimentation (since a moderate-intensity fire yields less than one-half as much sediment as would a severe fire).

If the young age class is 20 to 25 years, it would most likely be burned by the wildfire, there would be only a relatively small change in the wildfire effects, and there could be a net increase in sediment production relative to a single wildfire. (We assume that the supply of sediment is not quickly exhausted by a few fires.)

The Wildfire Front is of Comparable Size or is Small in Relation to the Younger Age Class. In these cases, there could be a substantial reduction in perimeter and area of the wildfire for those cases in which the young age class will not propagate fire or burning there is readily suppressed. Such was probably the case where the 1985 Wheeler Fire encountered under moderating weather the area of the 1971 Romero Fire (Salazar and González-Cabán 1987). The area so protected continues to age, and by continued mortality, to accumulate dead fuel. Depending on rates of fuel accumulation at later ages, which are not well established, the impact of a later wildfire there could be worsened if another prescribed fire does not intervene.

Use of prescribed fire is still limited in southern California, and although it has aided suppression of a few subsequent wildfires (Dougherty and Riggan 1982), it probably has not substantially altered the general impact of wildfires over the past decade. As we have discussed here, prescribed burning has potential for reducing some societal and environmental losses from wildfire, especially those associated with flooding, sedimentation, and water quality. But public policy for fire management remains driven by crises in funding rather than by the potential for averting catastrophic losses. The potential for fire management will be realized only when we are capable of monitoring and predicting long-term change in wildfire risks and articulating the benefits of an appropriate level of intervention. The most limiting factors in this analysis at present are the uncertainties regarding the propensity of young age classes to carry fire, the regional distribution of consumable biomass, and the rate of fuel accumulation in stands older than 3 decades.

How Predictable Will Be the Response to Climate Change?

Current predictions are that "business as usual" on earth will cause global mean air temperature to warm by as much as 2 to 5°C over the next century as a result of rising concentrations of radiation-absorbing trace gases in the atmosphere (Houghton et al. 1990). Regional predictions of potential climate change are highly uncertain, especially with regard to climate extremes such as those that most affect fires in southern California: Incidence of Santa Ana winds in autumn; heat waves associated with subtropical high pressure in midsummer; frequency of high rainfall years followed by drought; length and depth of multiannual

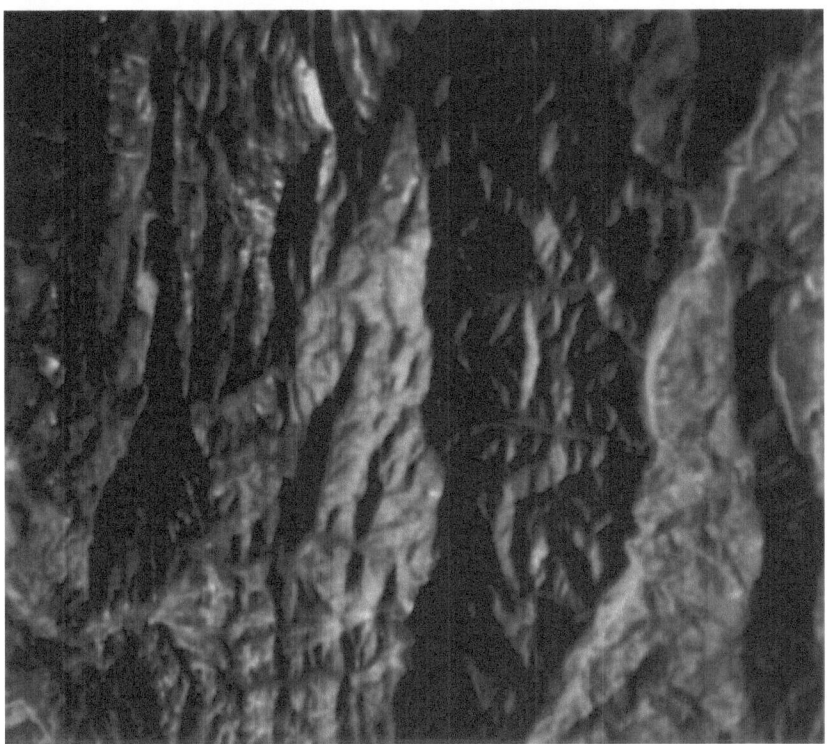

Figure 8–7. False-color composite image of the 1983 Matilija Fire (orange tones in center) within the perimeter of the 1985 Wheeler Fire. The 2-year-old-age class remained unburned despite extreme fire weather. The image was constructed from radiances measured at 0.91 to 1.05 (red), 0.63 to 0.69 (green), and 0.52 to 0.60 μm wavelength (blue) by an ADDS 1268 spectrophotometer (Daedalus Corporation). Data were collected 16 October 1985 aboard a NASA high-altitude aircraft.

droughts; and possibly, episodes of high temperatures, high humidity, and intermittent rain in midsummer.

Several of these conditions are now predicted with considerable uncertainty even in short-range forecasts and may remain unpredictable from long-range global climate models. One indicator of the adiabatically-heated and high-velocity Santa Ana winds, for example, is the development of a 4-mbar or greater offshore pressure gradient between Los Angeles and Tonopah, NV. Error in predicting the strength or relative location of the Great Basin high by a few hundred kilometers, and thereby the strength of this gradient and the Santa Ana winds, is fairly common within a 5-day forecast (N. Dean, National Weather Service, Fire Weather Forecasting Office, Riverside, CA, personal communication).

Annual precipitation and summer drought stress can be dominated by one or a few storm systems that may produce locally heavy or prolonged rainfall. Closely spaced storms during 9 successive days in February 1980 raised the total precipitation in the San Gabriel Mountains that season from average to twice the annual average and increased the area-specific stream flow to 55 cm from the expected average of 7.5 (Riggan et al. 1985). This recharged deeper soils and probably limited water stress during the succeeding summer. Rainfall of this magnitude in 1983 may have allowed an accumulation of leaf area that aggravated water stress condition in 1983–84 and caused the most severe dieback in decades. The 5-day forecasts are generally poor in predicting storm locations within 100 to 200 km or storm residence at one location, yet such large storms are probably one of the more important events in the ecosystem for a span of several years or even decades.

Even if global models could reliably predict average conditions for a season, we would have difficulty predicting extreme events in winds or rainfall, and these can dominate the ecosystem for years. Thus, by analogy, we find ourselves riding a rollercoaster in fog and not knowing if the amplitude of change is widening or subsiding. At present, the dips are catching us by surprise.

Conclusion

Without a doubt, southern California is an extremely challenging setting for natural resource management: Its environment is physically, biologically, and culturally dynamic. Management can affect the lives of millions of people, and these people exert a strong pressure on the environment. Catastrophic fires here will always be a threat to people and the environment. Climate extremes and impending climate change could drastically alter the situation in ways that would be very difficult to predict. But the impact of a fire regime can be altered through use of fire and a reduction of values at risk. To fail to act before the threat is immediate is to be continually responding to wildfires and paying their escalating price.

Note. Trade names, commercial products, and enterprises are mentioned solely for information. No endorsement by the U.S. Government is implied. This article was written and prepared by U.S. Government employees on official time, and it is, therefore, in the public domain and not subject to copyright.

References

Albini, F. 1980. *Thermochemical Properties of Flame Gases from Fine Wildland Fuels.* Intermountain Forest and Range Experiment Station, Forest Service, United States Department of Agriculture, Ogden, UT, Research Paper INT–243.

Brooks, F., Ferrin, D. 1991. Interim report for Cooperative Agreement PSW–90–045CA. On file, USDA Forest Service, Pacific Southwest Forest and Range Experiment Station, Riverside, CA.

Crist, C.R., Schoeneweiss, D.F. 1975. The influence of controlled stresses on susceptibility of European white birch stems to attack by *Botryosphaeria dothidea. Phytopathology* 65:369–373.

DeBano, L.F. 1981. Water repellent soils: A state-of-the-art, Pacific Southwest Forest and Range Experiment Station, Forest Service, United States Department of Agriculture, Berkeley, CA, General Technical Report PSW–46.

Dougherty, R., Riggan, P.J. 1982. Operational use of prescribed fire in southern California chaparral, pp. 502–510. *In* C.E. Conrad and W.C. Oechel (eds.), *Proceedings of the Symposium on Dynamics and Management of Mediterranean-Type Ecosystems.* Pacific Southwest Forest and Range Experiment Station, Forest Service, United States Department of Agriculture, Berkeley, CA, General Technical Report PSW–58.

Eaton, E.C. 1935. Flood and erosion control problems and their solution. *Proceedings of the American Society of Civil Engineering* 61:1021–1049.

Federal Emergency Management Agency. 1990. State and Federal hazard mitigation survey team report for the June 26–30, 1990 fires in southern California. FEMA–872–DR–CA.

Fendell, F.E., Carrier, G.F., Wolff, M.F. 1990. Wind-aided firespread across arrays of discrete fuel elements. Technical Report DNA–TR–89–193, Defense Nuclear Agency, Alexandria, VA.

Hanes, T.L. 1971. Succession after fire in the chaparral of southern California. *Ecological Monographs* 41(1):27–52.

Hickman, J.C. (ed.). 1993. *The Jepson Manual: Higher Plants of California.* University of California Press, Berkeley, CA.

Hinrichsen, D. 1987. The forest decline enigma. *Bioscience* 37:542–546.

Hodges, C.S. 1983. Pine mortality in Hawaii associated with *Botryosphaeria dothidea. Plant Disease* 67:555–556.

Horton, J.S. 1941. The sample plot as a method of quantitative analysis of chaparral vegetation in southern California. *Ecology* 22:457–468.

Houghton, J.T., Jenkins, G.J., Ephraums, J.J. (eds.). 1990. *Climate Change. The IPCC Scientific Assessment.* Cambridge University Press: Cambridge.

Jacks, P.M. 1984. The drought tolerance of *Adenostoma fasciculatum* and *Ceanothus crassifolius* seedlings and vegetation change in the San Gabriel Chaparral. Unpublished M.S. thesis, San Diego State University.

Kittredge, J. 1955. Litter and forest floor of the chaparral in parts of the San Dimas Experimental Forest, California. *Hilgardia* 23:563–596.

Marion, G.M., Black, C.H. 1988. Potentially available nitrogen and phosphorus along a chaparral fire cycle chronosequence. *Soil Science Society of America Journal* 52:1155–1162.

Paysen, T.E., Cohen, J.D. 1990. Chamise chaparral dead fuel fraction is not reliably predicted by age. *Western Journal of Applied Forestry* 5:127–131.

Pirsko, A.R., Green, L.R. 1967. Record low fuel moisture follows drought in southern California. *Journal of Forestry* 65:642–643.

Riggan, P.J., Goode, S., Jacks, P.M., Lockwood, R.N. 1988. Interactions of fire and community development in chaparral of southern California. *Ecological Monographs* 58:155–176.

Riggan, P.J., Lockwood, R.N., Lopez, E.N. 1985. Deposition and processing of airborne nitrogen pollutants in Mediterranean-type ecosystems of southern California. *Environmental Science and Technology* 19:781–789.

Riggan, P.J., Weirich, F.H., DeBano, L.F., Jacks, P.M., Lockwood, R.N., Colver, C., Brass, J.A. 1994a. Effects of fire severity on nitrate mobilization in watersheds subject to chronic atmospheric deposition. *Environmental Science and Technology* 28:369–375.

Riggan, P.J., Weirich, F., Lockwood, R.N., Jacks, P.M. 1994b. Debris flow generation after severe and moderate-intensity fires in southern California chaparral. Water Resources Research.

Rothermel, R.C. 1972. A mathematical model for predicting fire spread in wildland fuels. Intermountain Forest and Range Experiment Station, Forest Service, United States Department of Agriculture, Ogden, UT, Research Paper INT–115.

Rundel, P.W., Parsons, D.J. 1979. Structural changes in chamise (*Adenostoma fasciculatum*) along a fire-induced age gradient. *Journal of Range Management* 32:462–486.

Salazar, L.A., González-Cabán, A. 1987. Spatial relationships of a wildfire, fuelbreaks, and recently burned areas. *Western Journal of Applied Forestry* 2:55–58.

Sampson, A.W. 1944. Plant succession on burned chaparral lands in northern California. *California Agricultural Experiment Station Bulletin* No. 685.

Schlesinger, W.H., Gill, D.S. 1980. Biomass, production, and changes in the availability of light, water, and nutrients during the development of pure stands of the chaparral shrub, *Ceanothus megacarpus*, after fire. *Ecology* 61:781–789.

Schoeneweiss, D.F. 1975. Predisposition, stress, and plant disease. *Annual Review of Phytopathology* 13:193–211.

Schoeneweiss, D.F. 1979. Protection against stress predisposition to *Botryosphaeria* canker in containerized *Cornus stolonifera* by soil injection with Benomyl. *Plant Disease Reporter* 63:896–900.

Schoeneweiss, D.F. 1981. The role of environmental stress in diseases of woody plants. *Plant Disease* 65:308–314.

Schreiber, L.R. 1964. Stem canker and dieback of *Rhododendron* caused by *Botryosphaeria ribis*, Gross. and Dug. *Plant Disease Reporter* 48:207–210.

Shearer, B.L., Tippett, J.T., Bartle, J.R. 1987. *Botryosphaeria ribis* infection associated with death of *Eucalyptus radiata* in species selection trials. *Plant Disease* 71:140–145.

Specht, R.L. 1969. A comparison of the sclerophyllous vegetation characteristic of Mediterranean type climates in France, California, and southern Australia. II. Dry matter, energy, and nutrient accumulation. *Australian Journal of Botany* 17:293–308.

Spiers, A.G. 1977. *Botryosphaeria dothidea* infection of *Salix* species in New Zealand. *Plant Disease Reporter* 61:664–667.

Weirich, F.H. 1988. Field evidence for hydraulic jumps in subaqueous sediment gravity flows. *Nature* 332:626–629.

Wells, W.G.II. 1981. Some effects of brushfires on erosion processes in coastal southern California. *In Erosion and sediment transport in Pacific rim steeplands*. *IAHS Publication* 132:305–342.

Wilson, R.A. Jr. 1990. Reexamination of Rothermel's fire spread equations in no-wind and no-slope conditions. Intermountain Research Station, Forest Service, United States Department of Agriculture, Ogden, UT, Research Paper INT–434.

Worrall, J.J., Correll, J.C., McCain, A.H. 1986. Pathogenicity and teleomorphanamorph connection of *Botryosphaeria dothidea* on *Sequoiadendron giganteum* and *Sequoia sempervirens*. *Plant Disease* 70:757–759.

9.   The Role of Fire and Its Management in the Conservation of Mediterranean Ecosystems and Landscapes

Zev Naveh

The role of fire in conservation has to be considered within the context of the general goals of conservation in the Mediterranean. In recent years these have been broadened from the protection of threatened species in restricted areas to the conservation of biological plant and animal diversity and the ecological diversity of ecosystems and their ecological processes in the open landscape as a whole. As described elsewhere in more detail (Ricklefs et al. 1984), these processes include biogeochemical and hydrological cycling, primary and secondary production and the flow of energy and material sustaining them, mineralization of organic matter and, in general, all vital storage, transport, and regulation processes.

Most if not all ecological systems undergo cycles of disturbances and recovery that occur on characteristic scales of space and time. The regular occurrence of such disturbances – or perturbations – and the recovery processes following them may be necessary to maintain local ecological processes. They are thus also essential for the conservation of biological and ecological diversity in space and time. As will be explained further, we may even deal with "perturbation-dependent" systems (Vogl 1980), as in the case of the role of fire and other defoliation disturbances in Mediterranean sclerophyll forests and shrublands.

Until very recently, the role of fire in nature and land conservation in the Mediterranean has been treated in an entirely irrational and unscientific manner because of the destructive combination of fire with uncontrolled grazing and other abuses from which Mediterranean uplands have suffered for many centuries. Judged like goat grazing, only from its ill effects when abused, fire has been regarded in general as a wholly condemnable element that has to be prevented at all costs and at all times. All prominent Mediterranean phytosociologists mentioned it only in connection with human-induced regression stages, leading away from the so-called maquis-climax (Braun-Blanquet 1925; Kuhnholtz-Lordat 1939; Zohary 1962), and being thus detrimental to nature conservation (Tomaselli 1977).

The eminent ecologist Walter (1968), in his monumental study of the vegetation of the earth, was probably the first to recognize the importance of fire for the regeneration of *Pinus halepensis* and other conifers with a dense maquis shrub understory. He stressed its importance as one of the major ecological factors that shaped the Mediterranean landscapes into their present mosaic like regeneration and degradation patterns.

Full comprehension of the true role of fire in these landscapes can be reached only by recognizing its significance in the evolution of the Mediterranean flora and in the shaping of its vegetation patterns. These subjects have been discussed in more detail elsewhere (Naveh 1973, 1975, 1984, 1990a, 1991a; Naveh and Kutiel 1990), but I address them here only briefly. It is important to realize that the role of fire in the Pleistocene was not restricted only to the selection of fire-avoiding or fire-resistant genotypes and species. It had a decisive influence on the evolution of fire-induced and diversified landscapes. Through its use by the Paleolithic food gatherer and the Neolithic food producer it became also an important trigger in the cultural evolution of *Homo erectus* into *Homo sapiens* and thereby served as a major driving force in the coevolution of Mediterranean humans and their landscapes. During historical times fire became an important tool for pasture improvement by opening dense shrublands and increasing the herbaceous forage plants. In modern times uncontrolled wildfires, especially if combined with overgrazing, became a major cause for neotechnological landscape degradation. Its impacts on Mediterranean vegetation has been described by Le Houerou (1981) and Trabaud (1981).

In our study of fire, we have to consider both its detrimental effects and beneficial effects at all levels of the ecological and perceptional hierarchy and learn how to employ it in a judicious way as a cheap and efficient tool for the prevention of destructive wildfires and the conservation of the biological diversity and attractiveness of our open landscapes. This requires, above all, a more balanced, scientific approach to fire and its systematic study, without prejudice – like any other ecological disturbance factor – as part of multifactorial landscape functions.

Fire also plays a role as an integral part of the periodic human-induced perturbations and ecological processes driving Mediterranean uplands.

Some theoretical and practical conclusions for the role of fire in dynamic conservation management can be drawn from these studies.

Fire as Part of Multifactorial Landscape Functions

The effects of fire on vegetation and soil are very complex, not only because of the great complexity of Mediterranean plant communities and the interference of grazing and cutting with burning, but also because of the different responses to type and intensity of fire, its season and its frequency. Therefore, in order to avoid misleading, sweeping generalizations we have to distinguish, first of all, between wildfires and controlled fires.

But in addition, we have to take into consideration a great number of factors: For example, the fire frequency, intensity and extent; the ecological site conditions; its vegetation and fuel; the meteorological conditions prevailing during the fire and in the following seasons; and last, but not least, the biotic history of the burned site before and after the fire. These factors cannot be derived from superficial, circumstantial, and many times even biased observations. Nor can they be derived from short-term experimental studies carried out in a restricted area, not representing the landscape heterogeneity and the physical, biological, and cultural forces interacting with these fire effects.

The Combined Effects of Fire and Grazing on Vegetation and Soil Stability

As mentioned earlier, the effects of fire in this region have always been judged not on their own rights but in combination with grazing. Until very recently there were very few sclerophyll forests, woodlands, and maquis shrublands that were not grazed by cattle, sheep, and/or goats before and immediately after the fire.

All our comparative studies of fire effects of protected (as opposed to grazed) sites in different vegetation and soil types in Northern Israel (Naveh 1960, 1973, 1974) showed that the greatest damage was inflicted on vegetation and soil by grazing in the first winter and spring after the fire. The grazing of the young and lush leaves and twigs of the woody resprouters and of the seedlings and root-regenerating perennial grasses retarded their growth and regeneration. This favored the unpalatable, aggressive competitors of aromatic chamaephytes, especially *Labiatae* and *Cistus* species, which undergo fire-stimulated population explosion. These selective postfire grazing pressures have deflected the "auto succession" of the regenerating vegetation in the direction of the least palatable and most aggressive pyrophytes. This has turned large areas in the Mediterranean uplands into degraded and worthless scrubland.

The grazing history before and after the fire can have also detrimental effects on the degrees of postfire water runoff and soil erosion in the less fertile and more erodable soils, such as the highly calcareous pale rendzinas, especially if these soils have been disturbed and compacted prior to burning by uncontrolled grazing. However, even here, under well-distributed rainfall of 600 mm and more, if livestock grazing is postponed until the second spring after fire, the rapidly regenerating woody and herbaceous vegetation can ensure sufficient soil protection to prevent further degradation.

On the other hand, on much more fertile, well-developed, humus-rich terra rossa and brown rendzinas, covered by dense sclerophyll maquis shrubland before the fire, we could not observe any traces of water runoff or signs of soil splashing, creation of rills, and soil movement, on slopes of 30 to 40%, even after hot fires (Naveh 1973). These soils have a well-developed, humus-rich profile and even after high intensity rainfalls, the raindrops seep immediately into the soil, and the water running above the bare rock outcrops is intercepted close to the rock edges without creating any appreciable soil movement. These soils, containing more than 50% clay particles and about 12% organic matter in the upper 20 cm layer, have an excellent, stable granular structure that is apparently not impaired by the fire.

As can be seen in Table 9–1, even after losing about a fifth of the organic matter, 13 to 16% are still retained in the upper 4-cm soil profile beneath the burned sclerophyll shrubs and trees. This layer has been incinerated by the hot fire together with the conversion of the semi-decomposed litter and humus into a compact ash layer, underlain by a dark, charred soil layer.

Table 9–1. The Effect of Burning on Organic Matter of the Upper 4-cm Layer of Dark Brown Rendzina Maquis Soil Beneath Several Sclerophyllous Trees and Shrubs (Mazuba, Western Galilee, Israel, 1953)

Species	Soil Depth (cm)	Organic Matter (%)	
		Prefire	Postfire
Quercus calliprinos Webb.	0–2	18.1	6.4
Ceratonia siliqua L.	2–4	16.2	13.3
	0–2	23.6	14.6
Pistacia Palaestina Boiss.	2–4	18.3	12.3
	0–2	18.3	14.6
Pistacia lentiscus L.	2–4	16.5	12.3
	0–2	22.3	20.2
Average	2–4	16.3	14.2
	0–2	20.7	16.7
	2–4	17.0	13.2

Effects of Fire on Runoff and Sediment Loads from Small Plots on Mt. Carmel

In recent years, quantitative measurements of postfire runoff and sediment loads have been carried out in small plots on Mt. Carmel. Their results show clearly that their rates are determined by a great number of variables that need careful study and consideration from case to case and from site to site.

Thus after a recent, moderate wildfire that burned part of a planted, mature *Pinus halepensis* and *P. brutia* forest on Mt. Carmel, Kutiel and Inbar (1992) measured even higher total runoff and sediment discharge from the unburned plots. Zohar et al. (1990) and Lavee et al. (1991) observed similar phenomena. This could be explained as the result of the postfire increase in surface roughness and the creation of small depressions, the conversion of burned stumps into collecting undrained depression channels, promoting higher infiltration capacity, and the mulching effect of burned woody material and fine ash. On the other hand, after a recent very hot and destructive wildfire of dense maquis and mixed oak and pine forest on Mt. Carmel, increased runoff and sediment loads were measured in the first year on poor calcareous soils on steep slopes. This occurred mainly where the dead trees stumps were removed after the fire and the soil was disturbed and denuded. The runoff intensity and amount differed greatly from plot to plot and slope to slope. But in all cases, these rates were reduced to insignificant amounts in the second year after the fire, because of the rapid resprouting of the sclerophyll plants and the colonization of open space by herbaceous plants.

In general, such measurements of runoff and soil erosion carried out in small plots should be judged with great caution. Because of the great micro- and macrosite heterogeneity of these rocky slopes, both in physical and biotic conditions, they may have little meaning for the actual postfire events on the larger-scale landscape occurring in whole watershed and catchment processes.

In such a whole catchment area of the Rio Verde catchment in Southern Spain, May (1990) found a considerable increase of surface runoff immediately after the first hot wildfire, but in spite of relatively unfavorable conditions for vegetation recovery, hydrological behavior reverted to prefire conditions after three to four years. After two other fires no clearly detectable increase of surface runoff could be noticed. He concluded that it may be more useful to protect existing sclerophyllous, rapidly regenerating maquis in the lower part of the catchment, and *Pinus pinaster* forests in the upper zones vegetation from wildfires, including the use of prescribed burning, than to attempt pine forest reforestations. These are, in some cases, suspected of altering soil properties by trampling, thus impairing the vegetation's natural regeneration capacity. This is true also elsewhere, when the natural vegetation canopy is removed to enable the establishment of highly flammable dense pine and eucaiyptus stands.

Effects of Fire on Soil Fertility and on Herbaceous Plants

Another misleading generalization is that all these fires deprive the soil of its nutrients, especially nitrogen and phosphorous. This may be the case after very frequent fires, such as those studied by Trabaud (1983) in southern France, or after very hot wildfires. However, our recent studies on Mt. Carmel (Kutiel and Naveh 1987a, 1987b; Kutiel et al. 1990), as well as those by Kutiel and Shaviv (1989, 1992), have shown that the opposite is true. These results corroborated my earlier assumption that, as in comparable conditions in the California chaparral, also in the Mediterranean maquis and forests fire could have beneficial effects on nutrient cycling by mobilizing the nutrients tied up in the highly lignified wood and the slowly decomposing litter and duff accumulating in the A_{oo} and A_o profiles of the forest floor (Naveh 1967).

As shown in Figures 9–1 and 9–2, we found a striking increase in water-soluble nutrients, including nitrogen and phosphate in the first winter and spring after the fire in the upper centimeters of the soil, in spite of the loss of total nitrogen. This rather short-term postfire nutrient flush could be used by herbaceous fire followers for proliferous forage and seed production. They thus serve as an important link in the recycling of these nutrients to the soil from which the resprouting, deeper-rooted woody plants can benefit in the following years.

As shown in Table 9–2, the upper 2-cm layer of the burned brown rendzina soil, collected 2 months after a hot wildfire in 1983 on Mt. Carmel, produced a 6 times greater phytomass and 12 times more seeds in pot grown wheat. Of special significance for the enhancement of nutrient cycling is the 4.5 times greater root production, facilitating the manifold increase in nutrient accumulation in the plants. The striking postfire increase in seed production leads strong support to the hypothesis that such favorable ash seedbeds served as triggers for domestication of cereals and their incipient cultivation in slash-burn rotations in the early stages of the Neolithic agricultural revolution in the Levant (Naveh 1984).

However, neither this postfire mineral flush nor the temporary grass proliferation should be taken for granted. These nutrient levels depend not only on initial soil properties and the prefire vegetation but also on fire temperatures and its effects on complex microbiological processes in the soil (Kutiel and Shaviv 1989, 1992).

Figure 9–1. Changes in total nitrogen (A), ammonium (B), and nitrate (C) concentrations at the soil surface (0–3 cm) after a wildfire in a *Pinus halepensis* forest, Mt. Carmel, Israel. (PB, burned soil; PUB, unburned soil). (After Kutiel et al. 1990.)

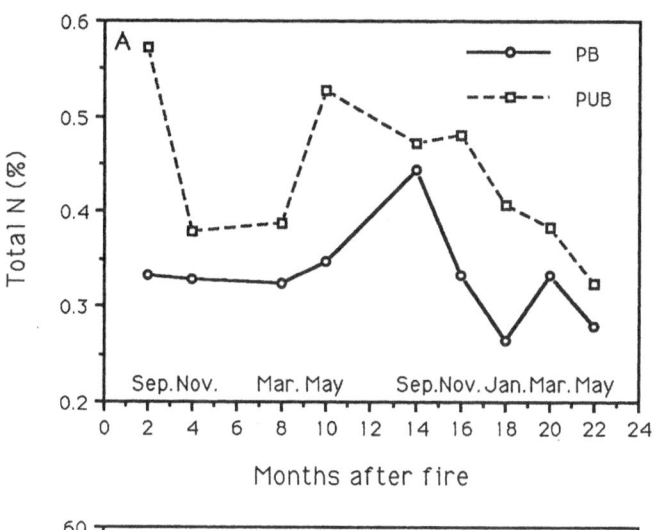

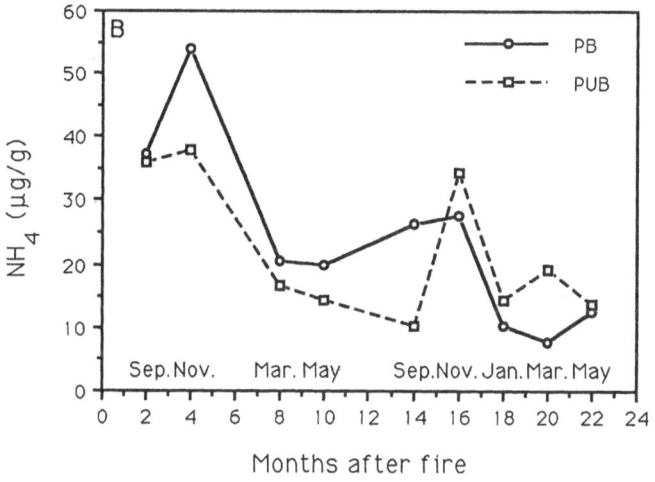

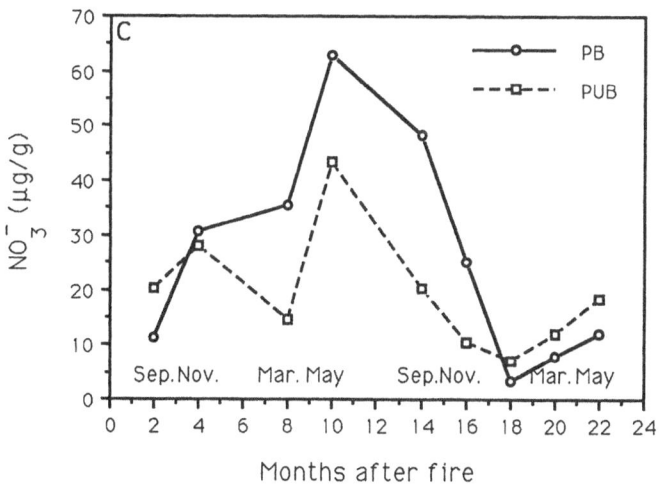

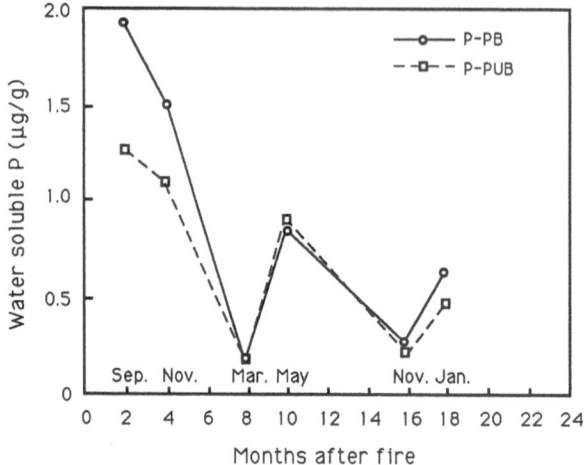

Figure 9-2. Changes in water-soluble phosphorus concentrations at the soil surface (0-3 cm) after a wildfire in a *Pinus halepensis* forest, Mt. Carmel, Israel. (PB, burned soil; PUB, unburned soil). (After Kutiel et al. 1990.)

Another highly stochastic factor is the possibility of the presence of phytotoxic and antibiotic agents in the litter and duff of aromatic *Labiatae*, *Cistus* and other species that are, at least partly destroyed by the fire, and others that may evolve after the fire. But much more systematic study is needed for understanding their role and interaction with fire, as well as of the fire-induced changes in soil biota and in the symbiotic soil–plant relations.

Table 9-2. Production and Nutrient Accumulation of Wheat Plants Grown in the Upper Soil Layer of Burned and Unburned Brown Rendina Pine Forest Soil of Mt. Carmel, Israel

	Burned (B)	Unburned (UB)	Ratio (B/UB)
Dry weight, shoots (g/m^2)	1142	197	5.3
Dry weight, roots (g/m^2)	709	158	4.5
Shoot-to-root ratio	2.2	1.2	1.8
Dry weight, spikes (g/m^2)	512	39	13.0
No. of seeds m^{-2}	13740	1102	12.5
Seed weight (g/m^2)	128	11	11.6
Nutrient accumulation N	0.6	0.2	3.0
Nutrient accumulation P	0.2	0.02	10.3
Nutrient accumulation Mg	0.6	0.08	7.5
Nutrient accumulation Ca	0.9	0.1	9.0
Nutrient accumulation Zn	0.02	0.005	4.0
Nutrient accumulation Fe	0.3	0.06	5.0

Note. (After Kutiel and Naveh, 1987).

In contrast to the almost 100% root regeneration of all Mediterranean sclerophylls, (if fires are not repeated in a highly artificial experimental manner year after year in fire ecology studies), the rate and extent of the postfire herbaceous plant colonization is highly stochastic. Therefore, no sweeping generalization should be made about the effect of fire on plant species diversity. As our studies showed, this is determined chiefly by the few perennial herbaceous plants that survive in the dense tree and shrub cover as shade-tolerant relicts, by the availability of vital seeds from these and other invading plants, and by the climatic conditions prevailing in the first and second rainy season after the fire. Important seed sources are small and patchy grass openings and edge habitats as well as waste heaps near human habitations. As will be shown later, the rates of recolonization also depend greatly on the size of the burned area and the intervals between fires at the burned site.

The vigorous herbaceous postfire flush is only temporary, depending on the rates of the recovery of the woody upper vegetation layer, but it has a great ecological and evolutionary significance. We can assume that the first opening of gaps in the closed tree cover of the pristine forests were caused in the Mediterranean by Paleolithic food gatherer–hunters in their search for food, fuel, and habitation and by the trampling of paths and digging for bulbs and earth animals. As documented by many archeological findings, the early use of fire by later Acheulian and Mousterian cultures in south France, Greece, Spain, and Israel had already started more than 500,000 years ago. It coincided with the widespread occurrence of wildfires from volcanic activity and lightning. Fire apparently served as the first extrasomatal source of energy for heating and cooking and for opening of dense forests and brush thickets. It thereby created more accessible and richer ecotones to facilitate hunting and food collecting. By encouraging the lush regenerating trees, shrubs, invading grasses, bulbs, and tuberous plants fire increased edible human and animal food. At the same time, this may have caused also the spreading of herbaceous fire-followers far beyond the previously mentioned human- and fire-induced forest gaps, grassy patches, and regeneration niches. In this way, the unique combination of human disturbances, the disposing of kitchen waste, and the recurring wildfires may have created favorable conditions for all those herbaceous colonizers that evolved in the Pleistocene and that could take best advantage of the improved light, moisture and fertility regimes.

In fact, in such forest gaps near waste heaps on Mt. Carmel we found that the most prolific fire followers were *Hordeum spontaneum*, the progenitor of domesticated barley, and *Piptatherum miliaceum*, the most abundant perennial grass whose plentiful, milletlike seeds can be baked and used as staple food. The same is probably true also for other grasses including *Triticum dicoccoides*, the most important progenitor of wheat, as well as many legumes that were domesticated later on. Although the slopes of the Eastern Galilee mountains facing the Jordan valley are burned almost every summer, all these plants recolonize again in the following winter rain season (Naveh 1973). Fire thereby plays a vital role in ensuring the genetic and floristic pool of many herbaceous plants, some of them even obligatory fire followers.

Thus, of great importance for the conservation of biological diversity and scenic attractiveness is the postfire increase of flowering geophytes, like *Bellevalia*, *Ornithogallum*, *Narcissus*, and *Cyclamen* that we observed in most of our studies in western Galilee and Mt. Carmel in Israel. Some of the orchid species, such as *Serapias vomeracea*, *Orchis papilionacens*, and *Ophris sintenissi* can be found only in well-lighted niches and not at all under the adjacent closed-brush canopy. If the increased postfire radiation intensity were the only ecological factor, we could ensure the survival of these heliophytic geophytes eventually by mechanical clearing and thinning of the dense maqui canopy and forest understory. But if fire also has direct, stimulative physiological effects on their regeneration and germination, as is the case with the earlier-mentioned *Piptatherum miliaceum* and other pyrophytes, then fire should become an indispensable tool in the dynamic conservation management of protected maquis and shrublands. At the same time, however, the regeneration of certain very fire-sensitive geophytes and other rarer species could be inhibited by too hot and too frequent fires. Therefore, this problem deserves careful study.

The Importance of a Hierarchical Approach to Fire Effects

It is important to realize that in our evaluation of all these fire effects we cannot blindly infer information from one level of the ecological hierarchy to the other. They have to be judged according to the hierarchical window through which they are viewed. If we look only at a single tree in a pine forest, then the fire means its death. And if we look at the fire in the first winter and see only the black, charred trees and ashes, we perceive it as an entirely catastrophic event. But if we enlarge our hierarchical scales both in space and time, we realize that for these plant communities, fire may be the only way to ensure the rejuvenation of the aging pine population. For their ecosystems it may be the only, or at least, the most efficient way to ensure their long-term biological productivity, diversity, and nutrient cycling and for their landscapes as a whole to retain their spatio-temporal heterogeneity, attractiveness, and dynamic flow equilibrium in space and time. In this way, by broadening the scales from the individual plant to the landscape hierarchy level, fire has become incorporated into the system and should not be considered merely as a natural or anthropogenic perturbation. Additional examples of such "ecosystem incorporation" of fire have been described by O'Neill et al. (1986).

Of special importance is the consideration of the spatio-temporal scales on postfire species richness and survival. As mentioned earlier, in general, there is an increase in herbaceous species richness in the first years after the fire. But at the same time, it may also cause the disappearance of a few, fire-sensitive species. This was probably the case in a very hot wildfire that raged for 3 typical Sharav days with strong, hot, and dry winds in November 1983 on Mt. Carmel. It swept through 330 ha and hit most of the Nature Reserve, preserving the only large natural *Pinus halepensis* forest in Israel. Unfortunately, the policy of complete protection and noninterference led to the accumulation of great loads of dry fuel in the almost impenetrable forest understory. As no watering points, fire breaks and sufficient access roads were provided, the forest could not be saved.

As mentioned earlier, we have to consider also the temporal scales: If the interval between two fires is longer than the reproduction capacity of the burned plants, then the fire can lead to their extinction. In the case of *Pinus halepensis* this may be a hundred years, but in the case of the earlier-mentioned geophytes in the forest understory, it is only a couple of years. The same is true also for too-short fire frequencies that do not allow the plant sufficient time for reproduction and/or regeneration.

This temporal scale is interrelated with the spatial scale. The narrower the ratio between the area occupied by this species in the landscape and the burned area, the poorer its prospects for survival. But if the species are distributed over a wide range of habitats in a heterogeneous landscape, burning part of this system will not endanger the species, even if the regeneration in the burned part is prevented or is only very slow. It is, therefore, obvious that in order to ensure the highest attainable plant and animal species diversity with the help of fire, its management must be highly flexible in space and time. It must be adapted to the requirements and fire responses of target species to which these management strategies are aimed, and to the micro- and macro-site heterogeneity of the landscape as a whole. This can be realized by rotational fire management in space and time, in many ways similar to that practiced by pre-agricultural, indigenous cultures in comparable conditions in California and west Australia, and probably also by epipaleolithic gatherer–hunters, such as the Natufian in the Levant (Naveh 1984, 1990b).

The Need for Multifactorial Fire Functions

The major conclusion to be drawn from this discussion is the realization that all fire effects have to be treated as multifactorial ecosystem and landscape functions at differing spatial and temporal scales and rates. This can be done by using the functional–factorial approach of Jenny

(1961) and distinguishing between controlling state factors and their depending variables:

$$L_{s,v,a,ha,\ldots} = f(F_{t,se,\ldots}\,[H,P,R,C,O,\ldots T])$$

In this equation landscape (L) features, namely soils (s), vegetation (v) animals (a), human-made artifacts (ha), as well as other unspecified variables (...) are dependent postfire variables of the function of fire F and its parameters, and especially temperature(t), season (se) and other unspecified parameters (...). These fire parameters interact with other controlling state factors, chiefly human land use (H), soil parent material (P) relief (R), the flux potentials of climate (C) and organisms (O), in addition to unspecified factors (...) and the time (T) since the fire event, during which this function is measured.

An excellent example of such a multifactorial approach is a recent study of microscale vegetation patterns after controlled burning of California coastal chaparral by Davis et al. (1989). They used multivariate analysis to document scale-dependent correlation between preburn seed assemblages, preburn canopy cover, microtopography, soil temperatures during burning, postburn seed assemblages, and post-burn vegetation.

As long as fire ecologists in the Mediterranean are not allowed to use controlled burning for their studies, and have to rely only on wildfires, they will not be able to gain control of all of these factors. However, with the help of such multifactorial equations they can at least realize the most decisive driving forces of this fire landscape function that may change from case to case. Let us hope, however, that in the near future much more well-designed, prescribed, and controlled fires will enable a more efficient and reliable study of these functions and their impacts on the landscape.

Mediterranean Uplands as Perturbation-Dependent Homeorhetic Ecosystems

Throughout the closely interwoven natural and cultural processes that shaped these landscapes and their vegetation for many thousands of years, the seminatural and agricultural vegetation of open forests, shrublands, woodlands, grasslands, and terraces have all become part of closely interwoven landscape mosaics with highly dynamic degradation and regeneration vegetation patterns in time and space. The Bible has provided ample proof of the important role of natural and human-caused fires in early historical–pastoral land use in the Levant. Many classical Greco-Roman sources mention the beneficial effects of stubble burning and wood ashes as fertilizers and for weed and pest control, as well as for brush pasture improvement in forests and woodlands. Some of these applications of fire have continued until present times, together with grazing, cutting, coppicing, terracing, and patch cultivation in traditional agricultural practices.

It can be assumed that in the traditional Mediterranean pastoral systems, the great climatic seasonal and annual fluctuations induced changes in productivity, which acted as effective negative regulative feedbacks in preventing overgrazing, similar to the natural regulating of wildlife populations. At the same time, overcutting and overburning were prevented by burning and coppicing rotations necessary to ensure sustained productivity and sufficient recovery. These regular, centuries-lasting grazing, burning, and coppicing regimes and agro-pastoral functions led to the establishment of a human-maintained balance between the tree, shrub, herb, and grass layers in those forests, woodlands, and shrublands that were neither overgrazed nor overcoppiced or were completely rested for prolonged periods. For such situations, the term Perturbation Dependent Ecosystems, coined by Vogl (1980) seems to be very appropriate.

In combination with the great macro- and microsite heterogeneity of the rocky and rough terrain and the climatic annual and seasonal fluctuations, these periodic and mostly cyclic human perturbations induced the unique ecological and cultural landscape diversity of Mediterranean uplands and their earlier-mentioned dynamic vegetation patterns and processes. In most cases, these do not fit any preconceived, deterministic, relay-successional sequences. They also render thermodynamic ecosystem interpretations of Odum (1971) as doubtful if applied to these Mediterranean seminatural ecosystems. Since the system is not returned to a constant equilibrium, like in homeostatic climax systems, it returns to its perturbation trajectory and thereby is changing in the same way as it has in the past. For such a dynamic flow equilibrium, the eminent geneticist Waddington (1977) coined the term homeorhesis (from the Greek, meaning "preserving the flow"). In our seminatural Mediterranean landscapes, these changes are induced by human disturbances. They are, therefore, homeorhetic human perturbation-dependent systems, which have acquired long-term adaptive resilience and metastability (Naveh 1991b; Naveh and Lieberman 1993).

However, this dynamic flow equilibrium is currently threatened by the replacement of these traditional agro-pastoral functions by accelerating "neotechnological" degradation functions of intensified traditional and modern agro-pastoral pressures and/or land abandonment and negligence and their combination with neo-technological landscape degradation, and pollution, urban industrial sprawl, and uncontrolled mass recreation and tourism.

In addition to the loss of open landscapes and their fragmentation by urban-industrial developments, the floristic and structural diversity of the natural vegetation, as well as the faunistic species richness and abundance are constantly being reduced. This impoverishment is not only the result of the heavy human and livestock pressures in densely populated hill and mountain areas in the Levant and in North Africa, but also that of complete cessation of human interference, including burning in depopulated areas, as well as in protected nature reserves and parks. In many of these, however, heavy pressures of mass recreation are also threatening floristic and structural diversity. In general, after the decline of human and livestock pressures, the initially vigorous vegetative regeneration of sclerophylls from stunted shoots and almost imperceptible rootstocks is followed by the gradual encroachment of the shrub canopy and the almost total suppression of the herbaceous understory in undisturbed and protected maquis and forests. Noninterference is turning from a blessing to a curse when the denser brush thickets becomes stagnant and senescent and more and more prone to hot and devastating wildfires.

In Israel, such impenetrable and monotonous, one-layered, tall shrub climax communities are dominated almost exclusively by *Quercus calliprinos* with a few occasionally codominant shrubs and subordinate dwarf shrubs and very few, shade-tolerant perennial grasses and geophytes, which occur near open patches of rock outcrops and shrub edges. This results in the reduction in plant species richness of about 75% (to less than 30 species/0.1 ha), as well as significant reductions in other diversity parameters, and in richness and abundance of birds, rodents, reptiles, and insects (Naveh and Whittaker 1979). A survey of all nature reserves of northern Israel, in an area of about 30,000 ha confirmed our findings on a much larger scale: In protected and too lightly grazed sites species richness and diversity was not only lower, but also the frequency of wildfires was higher than in moderately and heavily grazed ones (Noy-Meir and Kaplan 1991).

The same undesirable tendencies toward lower structural and floristic diversity, combined with increased vulnerability to fire, can be observed in all other Mediterranean countries. Among others, these tendencies have been reported also in Adriatic *Q. ilex* forests (Horvat et al. 1974), as well as on the island of Lokrum, near Dubrovnic, which has been protected as a nature reserve since 1948 (Ilijanic and Hecimovic 1981). On larger landscape scales, these threatening trends have been reported from almost all Mediterranean countries: Farina (1991), González Bernáldez (1991), Vos and Stortelder (1992), and in a recent symposium on the future of Mediterranean landscapes at Montecatini (Farina and Naveh 1993).

As emphasized by Ruiz de la Torre (1985) in the important book on the conservation of Mediterranean plants, most endemics in the Mediterranean are small shrubs and herbs. They are, therefore, among the candidates to go extinct through brush re-encroachment, resulting from the conservation policies of total protection from human interference. Another important botanical group especially endangered due to its ornamental value are the flowering geophytes, which include many endemic and rare species. In most Mediterranean countries these are picked and sold and their bulbs exported in large quantities. In Israel, where they are better protected, this danger has been reduced. Even so, because of their higher light demands, they are smothered out by the cessation of burning, grazing, and cutting, as well as by afforestation.

One of the results of the disruption of the traditional agropastoral flow equilibrium is the vicious circle of increasing wildfire hazard, fed on positive, destabilizing runaway feedback couplings between the cessation of traditional defoliation pressures and the resulting accumulation of heavy fuel loads, planting of highly inflammable pine and eucalyptus forests, and exponentially growing urban and recreational pressures. Therefore, despite (and sometimes even because of) efforts and heavy expenditures on total fire suppression, wildfires in the Mediterranean uplands are becoming more frequent and as devastating as ever and are causing heavy damages each year, far exceeding the costs of fire and fuel management. At the same time, because of the total ban on burning in most Mediterranean countries, even its use for research purposes has been prohibited, as well as its application in a judicious way for the reduction of fuel loads and prevention of much hotter wildfires.

The predicted global climatic changes in rainfall and temperature regimes, resulting from the "greenhouse effect" could further aggravate these threats. If the rise in CO_2 concentrations could further enhance the brush encroachment in abandoned and neglected forests and maquis and their floristic impoverishment, their increased fuel loads will cause even hotter and more destructive wildfires. The anticipated rise in summer temperatures and in evaporation rates could have severe repercussions on the most mesic and richest plant communities, their diversity and stability, and could further aggravate the climatic fire hazard conditions. The rise in temperature may also increase further the combined desiccation effects of air pollutants, especially ozone, and pests on conifer species (Naveh et al. 1981) and thereby also increase even further the fire hazards of these highly flammable trees.

Some Theoretical and Practical Conclusions for Dynamic Conservation Management: Discussion

From the foregoing description, it is obvious that efficient measures are urgently needed for protection of the open landscapes and their great structural, floristic, and faunal diversity. These measures should not only prevent further traditional and modern abuses, but also ensure the specific defoliation pressures of burning, grazing, coppicing, cutting, and so forth under which they evolved. Most of these uplands are still distinguished by their great biological, ecological, and cultural diversity in space and time. But this is the case only where their vegetation has not yet been converted into dense, monotonous, and biologically almost sterile pine or eucalyptus forests, or has been depleted either by overuse into scrub and rock deserts or by underuse into monotonous, highly flammable forest and shrub thickets.

One of the main conclusions to be drawn is that conservation management strategies in the Mediterranean cannot be guided by deterministic and illusory succession-to-climax theories of noninterference. According to such theories, complete and prolonged protection from human "disturbances" – including also controlled burning – should enable the socalled "natural vegetation development toward stable maquis and forest climax communities" (Tomaselli 1977). On the contrary, we have to conserve and to reestablish their homeorhetic flow equilibrium by continuing or simulating all the ecological processes and defoliation pressures to which they have been adapted throughout their long cultural history.

The importance of the re-establishment of this multifactorial homeorhetic flow process by active and dynamic conservation management can be supported by recent insights into the thermodynamic behavior of self-organizing (or autotopoietic), dissipative structures. For this purpose, we have to distinguish between natural and seminatural ecosystems. In natural ecosystems, high-quality potential and chemical energy (and therefore low entropy-producing energy) is derived from solar energy and its conversion by photosynthetic and other biological production functions. Part of this energy is dissipated into low-quality metabolic heat and respiration, and thus negentropy as a measure of organizational order and information is built up in the landscape by structural and spatial heterogeneity, high species diversity and complexity in food chains webs. Simultaneously, entropy production – as a measure of homogeneity, and disorder – is also minimized by the protection and stabilization functions of the "living sponge" of vegetation cover. These reduce the rate of kinetic energy and heat flows and their destructive and destabilizing impacts on the landscape. Therefore, in such undisturbed and mature ecosystems, a steady state is presumably reached in which high negentropic order and information are maintained through homeostatic self-stabilization.

The thermodynamic behavior and homeorhetic self-regulation and self-organization of our seminatural Mediterranean sclerophyll forests, maquis, shrublands and woodlands are very different. As perturbation-dependent, nonequilibrium systems, they behave like autopoietic dissipative structures. According to Prigogine (1976), they create "order through fluctuation" and "order out of chaos" (Prigogine and Stenger 1984). Such dissipative structures are maintained and stabilized only by permanent energy/matter and entropy exchange processes. Driven by positive feedbacks of environmental and internal fluctuations, they move to new regimes that generate the conditions of renewal of higher entropy production.

In our case, such new regimes are apparently created by the periodical perturbations of fire, grazing, and cutting. But this will happen only if sufficient time has been allowed for there to be a generation phase and thereby also for the import of negentropy through intensified photosynthetic growth and generation processes. At the same time, the system can actually use free energy to reorganize itself with increasing structural complexity, biological diversity, and productivity. But if these perturbation cycles are too frequent and severe, external entropy exchange may become more and more positive and disorder remains at a high level. The same is also true, if these perturbations are stopped altogether either by total protection and noninterference or by abandonment. In this case negentropy and information rise in the early regeneration phase, but with the lack of further perturbations, the rates of entropy production again increase and disorder becomes more and more positive.

This is expressed by the monotony and low structural, floristic, and faunistic diversity of undisturbed Mediterranean forest and maquis thickets and by their high flammability. They are not comparable to the natural, rich, and stable "mature" climax ecosystems, described by Odum (1969), because they are aging and becoming more and more stagnant and senescent, more and more flammable, and thus unstable and species-poor through time.

This process can be illustrated with the help of Prigogine's dissipative function, in which entropy (s) and thus also disorder (D) grow at the rate ds/dt:

$$D = ds/t$$

D may be positive, negative, or zero. If it is zero, then the system is in a stationary state – as in the homeostatic "climax" state of natural systems. If it is positive ($D > 0$), then the system is in a state of progressive disorganization and conversely the rate of negentropy and information (info) decreases and it looses its capacity for self-stabilization and self-organization, as in the state of too-frequent or no perturbations:

$$D > 0 = dinfo/dt < 0$$

But if D is negative ($D > 0$), then the system is in a state of progressive organization and increases its negentropy and information, as in the case of optimum perturbations, when the homeorhetic metastability and thus the capacity of constant self-organization and stabilization can be maintained:

$$D < 0 = dinfo/dt < 0$$

Figure 9–3 illustrates the results of these 3 different perturbation regimes.

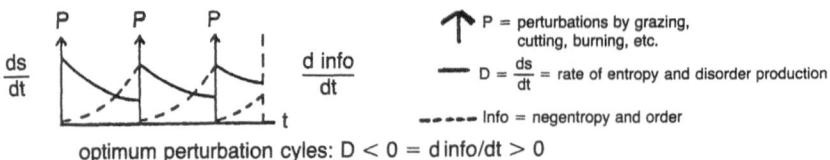

optimum perturbation cyles: D < 0 = d info/dt > 0

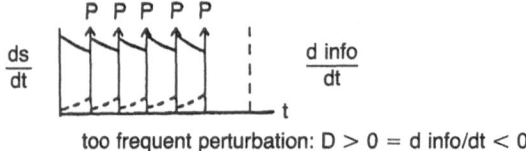

too frequent perturbation: D > 0 = d info/dt < 0

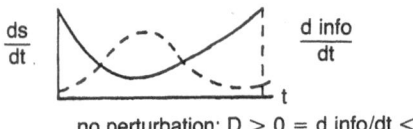

no perturbation: D > 0 = d info/dt < 0

Figure 9–3. The thermodynamic reactions of Mediterranean ecosystems to three perturbation regimes. (After Naveh 1987.)

The determination of management strategies for such "optimum" regimes of perturbation and enhancement of biological diversity alone (in nature reserves) or in combination with other land use goals, such as increase of economic production, recreation amenities, and scenic values, requires systematic, long-term studies on landscape scales. These studies will have to provide the answer if and when controlled burning, both for the reduction of fuel and prevention of destructive wildfires and for dynamic conservation management, can be replaced completely by chemical or mechanical means or by domestic or wild animal grazing.

Conclusion

The main practical conclusion from this discussion is the need for flexible management strategies, based on optimum regimes of perturbation and defoliation pressures, including controlled burning, as well as the conservation and restoration of all other vital ecological processes, aimed at the maintenance and enhancement of biological diversity and metastability (Naveh and Lieberman 1993).

This should be the major goal in nature reserves but in the open landscape it should be combined with other, multibeneficial land use goals, such as increase of economic production, recreation amenities, and scenic values. This requires systematic, long-term studies on ecosystem and landscape scales for which the previously described multifactorial functional approach can be very useful.

The results of our studies and most others conducted in recent years in the Mediterranean indicate that fire may be vital for the direct stimulation of germination, growth, regeneration, and rejuvenation of many species, for the removal of heat unstable phytotoxic agents in the unburned litter, and for efficient nutrient cycling.

This need for a change from total fire suppression to dynamic fire and fuel management as part of wildlife conservation has been supported also by the results of the studies of Walter (1977) on the indirect beneficial effects of controlled burning on bird populations in Sardinia and by Prodon (1992). This will require well-integrated fire ecology, ecosystem, and landscape ecology studies, aimed at dynamic conservation management, the reduction of fuel and prevention of destructive wildfires, in combination with controlled grazing and browsing of wild and domestic ungulates or other defoliation means. These studies should provide comprehensive and conclusive answers as to if, when, and how controlled burning should be applied.

In concluding an earlier review on the role of fire in the Mediterranean region, I stated, "Fire has apparently served as a most beneficial tool in the skilled hand of our Mediterranean ancestors but it is being now neglected and rejected. It is up to the ecologist and enlightened land user to show how fire can be turned from a curse in Mediterranean lands to a blessing" (Naveh 1974).

Fortunately, there are many encouraging signs that this challenge has already been accepted by many young and able ecologists in several Mediterranean countries. This was well reflected in the recent Third European Symposium on Fire Ecology (Goldammer 1990) and will, hopefully, lead in the near future also to practical results of more rational and efficient fire and fuel management in our uplands and their threatened biological and cultural diversity.

However, ultimately the success of such a new dynamic landscape and fire and fuel management policy within the broader framework of holistic landscape ecological masterplans will depend on the creation of awareness of the biological and cultural values of these landscapes and their total ecodiversity, concern about their future, and motivation for active involvement in their sustainable utilization, conservation and restoration. This will require a total shift from the still mostly prevailing one-sided instrumental and exploitative attitudes by an educational process which should reach all those who care for these landscapes, those who live from them, and those who deal with them at all levels of the decision making process. For this purpose much more efficient communication tools are necessary to bridge the gap between the academicians and the professionals, the conservation and restoration–minded ecologist and the production-minded foresters, agronomists, and economists, as well as between all these specialists and the public at large.

Such an important tool could be the Green Books (formerly called Red Books) for the conservation of threatened, highly valuable landscapes. As proposed by Naveh and Lieberman (1993), these Green Books should present in the local language(s) in clear nontechnical terms with ample maps and illustrations not only recent, adverse changes endangering both natural and cultural assets and scenic and economic values but also suggest alternative, sustainable land use strategies based on holistic landscape planning and dynamic conservation management. Being not only descriptive but also anticipatory, they could help to change the attitudes of politicians and decision makers, and provide practical guidance for holistic, sustainable and multibeneficial land use planning and management. Their object is not to produce another scientific report to be published and filed away sooner or later, but to become a usable document for professionals and the citizens. In this way, its semantic information will be transformed into pragmatic information, which becomes meaningful by its effects on the receiver and is expressed in his actions.

Such Green Book projects are presently initiated by the Working Group for Landscape Conservation of the IUCN – World Conservation Union – Commission on Environmental Strategies and Planning. With the help of advanced quantitative landscape ecological methods a uniform methodology will be developed, including integrated ecological and socio-economic field surveys, appraisals and landscape classification, combined with remote sensing and dynamic Intelligent Geographical Information Systems (IGIS), and with multivariate statistical analysis and clustering methods.

The first of such Green Book case studies was completed in 1993 by a multidisciplinary and multinational team from the University of Thessaloniki and the University of Cambridge in Western Crete (Grove et al. 1993). The detailed recommendations for sustainable land use strategies show how the demands to safeguard "soft" biological and cultural values could be reconciled with controlled utilization of "hard" values which are vital for socio-economic advancement of the local population.

References

Braun-Blanquet, J. 1925. Die *Brachypodium ramosum – Phlomis lychnitis –* Assoziation der Roterdeboeden Suedfrankreichs. *Festschrift C. Schroeder.* Veroeff. Geobot. Inst. Ruebel, H. 3.

Davis, F.W., Borchert, M.I., Odion, D.C. 1989. Establishment of microscale vegetation patterns in maritime chaparral after fire. *Vegetation* 84:53–67.

Farina, A. 1991. Recent changes of the mosaic patterns in a montane landscape (North Italy) and consequences on vertebrate fauna. Options Mediterranéennes 15:121–134.

Farina, A., Naveh, Z. (eds.). 1993. Landscape approach to regional planning. The future of Mediterranean landscapes. *Landscaping and Urban Planning* 24: special issue, July 1993.

Goldammer, J.G., Jenkins, M.J. (ed.). 1990. *Fire in Ecosystem Dynamics, Mediterranean and Northern Perspectives*. SPB Academic Publishing, The Hague, The Netherlands.

González Bernáldez, F. 1991. Ecological consequences of the abandonement of traditional land use systems in central Spain, pp. 23–29. *In* J. Baudry and R.G.H. Bunce (eds.), *Land Abandonement and Its Role in Conservation*. Proc. Zaragozsa Seminar 10–12 September 1989. Options Mediterraneennes. Ser. A15.

Grove, A.T., Ispikoidis, J., Kazaklis, A., Moody, J.A., Papanastasis, V., Rackham, O. 1993. *Threatened Mediterranean Landscapes: West Crete*. Final Report to EEC. Department of Geography, University of Cambridge.

Horvat, I., Glavac, V., Ellenberg, H. 1974. *Vegetation Suedosteuropas*. Gustav Fischer Verlag, Stuttgart, Germany.

Ilijanic, L.C., Hecimovic, S. 1981. Zur Sukzession der Mediterranean Vegetation auf der Insel Lokrum by Dubrovnic. *Vegetation* 46:75–81.

Jenny, H. 1961. Derivation of state factor equations of soil sand ecosystems. *Soil Sci Proc*. 385–388.

Kuhnholtz-Lordat, G. 1939. *La terre incendiée*. Edit. Mais Carrée, Nimes, France.

Kutiel, P., Naveh, Z. 1987a. Soil properties beneath *Pinus halepensis* and *Quercus calliprinos* trees on burned and unburned mixed forest on Mt. Carmel. Israel. Forest Ecology and Management 20:11–24.

Kutiel, P., Naveh, Z. 1987b. The effect of fire on nutrients in pine forest soil. *Plant and Soil* 104:262–274.

Kutiel, P., Shaviv, A. 1989. Changes of soil N–P status in laboratory-simulated forest fire. *Plant and Soil* 120:57–63.

Kutiel, P., Naveh, Z., Kutiel, T. 1990. The effect of wildfire on soil nutrients and vegetation in an Aleppo pine forest on Mt. Camel, Israel, pp. 85–94. *In* J.G. Goldammer (ed.), *3rd European Symposium on fire ecology, Freiburg University, 16–17 May 1989*. SPB Academic Publishing, The Hague, The Netherlands.

Kutiel, P., Shaviv, A. 1992. Effects of soil type, plant composition and leaching on soil nutrients following a simulated forest fire. Forest Ecology and Management 53:329–343.

Kutiel, P., Inbar, M. 1992. Fire impact on soil erosion in a Mediterranean pine forest plantation. *Catena*. 20:129–139.

Lavee, H., Benyamini, Y., Kutiel, P., Segev, M. 1991. Infiltration, runoff and erosion processes as influenced by forest fires in the Carmel Mountain, Israel. pp. 24–36. *European Society for Soil Conservation*. Conf. Soil Erosion and Degradation as a Consequence of Forest Fires. Barcelona and Valencia, Spain. (in press)

Le Houerou, H.N. 1981. Impact of Man and his animals on Mediterranean vegetation, pp. 479–522. *In* F. Di Castri, D.W. Goodall, and R.L. Specht (eds.), *Ecosystems of the World 11: Mediterranean-Type Shrublands*. Elsevier Scientific Publishing Company Amsterdam.

May, T. 1990. Vegetation development and surface runoff after fire in a catchment of southern Spain, pp. 117–126. *In* J.G. Goldammer and M.J. Jenkins (eds.), *Fire in Ecosystem Dynamics, Mediterranean and Northern Perspectives*. SPB Academic Publishing, The Hague, The Netherlands.

Naveh, Z. 1960. Agro-ecological aspects of brush range improvements in the maquis belt of Israel. Unpublished Ph.D. thesis, Hebrew University, Jerusalem.

Naveh, Z. 1967. Mediterranean ecosystems and vegetation types in California and Israel. *Ecology*, 48:445–459.

Naveh, Z. 1973. The ecology of fire in Israel, pp. 130–170. *Proc. 13th Ann. Tall Timbers Fire Ecol. Conf.*, Tallahassee, FA.

Naveh, Z. 1974. Effects of fire in the Mediterranean region, pp. 401–434. *In* T.T. Kozlowski and C.E. Ahlgren (eds.), *Fire and Ecosystems*. Academic Press, New York.

Naveh, Z. 1975. The evolutionary significance of fire in the Mediterranean region. *Vegetation* 9:199–206.

Naveh, Z. 1984. The vegetation of the Carmel and Nahal Sefunim and the evolution of the cultural landscape. *The Sefunim Prehistoric Sites Mount Carmel, Israel*, pp. 23–63. Ronen BAR International Series 2310, Oxford, England.

Naveh, Z. 1987. Biocybernetic and thermodynamic perspectives of Landscape functions and land use patterns. *Landscape Ecology* 1:75–83.

Naveh, Z. 1990a. Fire in the Mediterranean: A landscape ecological perspective, pp. 1–20. *In* J.G. Goldammer (ed.), *3rd European Symposium on Fire Ecology, Freiburg University 16–17 May 1989*. SPB Academic Publishing, The Hague, The Netherlands.

Naveh, Z. 1990b. Ancient man's impact on Mediterranean landscapes in Israel – ecological and evolutionary perspectives. pp. 43–50. *In* S. Bottema, G. Entjes-Nieborg, and W. Van Zeist (eds.), *Man's Role in the Shaping of the East Mediterranean Landscape*. AA Balkema, Rotterdam.

Naveh, Z. 1991a. The role of fire in Mediterranean vegetation. Botanika Chronica, 10:385–405.

Naveh, Z. 1991b. Mediterranean uplands as anthropogenic perturbation dependent systems and their dynamic conservation management, pp. 544–556. *In* O.A. Ravera (ed.), *Terrestrial and Aquatic Ecosystems, Perturbation and Recovery*. Ellis Horwood, New York.

Naveh, Z., Whittaker, R.H. 1979. Structural and floristic diversity of shrublands and woodlands in Northern Israel and other Mediterranean areas. *Vegetation* 41:71–190.

Naveh, Z., Steinberger, E., Chaim, S. 1981. Photochemical air pollutants – A new threat to Mediterranean conifer forests and upland ecosystems. *Environ. Conserve.* 7:301–309.

Naveh, Z., Kutiel, P. 1990. Changes in the Mediterranean vegetation of Israel in response to human habitation and land use, pp. 259–300. *In* G.M. Woodwell (ed.), *The Earth in Transition: Patterns and Processes of Biotic Impoverishment*. Cambridge University Press, Cambridge, England.

Naveh, Z., Lieberman, A.S. 1993. *Landscape Ecology Theory and Applications*. Springer-Verlag, New York. Second Edition.

Noy-Meir, I., Kaplan, D. 1991. The affect of grazing on the herbacious Mediterranean vegetation and its implications on the management of nature reserves. Internation Report of the Nature Conservation Authorities of Israel, Jerusalem (Hebrew).

Odum, E.P. 1969. The strategy of ecosystem development. Science 164:262–270.

Odum, E.P. 1971. *Fundamentals of Ecology*, 3rd ed. W.B. Saunders, Philadelphia.

O'Neill, R.V., DeAngelis, D.L., Waide, J.B., Allen, T.E.H. 1986. *A Hierarchical Concept of Ecosystems*. Princeton Press, NJ.

Prigogine, I. 1976. Order through fluctuations: Self-organization and social systems, pp. 93–130. *In* E. Jantsch and C.W. Waddington (eds.), *Evolution and Consciousness: Human Systems in Transition*. Addison-Wesley, Reading, MA.

Prigogine, I., Stengers, I. 1984. *Order Out of Chaos*. Bantam Books, New York.

Prodon, R. 1992. Animal communities and vegetation dynamics: Measuring and modeling animal community dynamics along forest succession, pp. 23–29. *In* A. Teller, P. Mathy, and J.N.R. Tellers (eds.), *Responses of Forest Ecosystems to Environmental Changes. Proc. First Europe. Symp. on Terrestrial Ecosystems, Forest and Woodlands.* Elsevier, London.

Ricklefs, R.E., Naveh, Z., Turner, R.E. 1984. Conservation of Ecological Processes, IUCN Commission on Ecology, paper number 8. *The Environmentalist* 4:1–16. (Supplement)

Ruiz de la Torre, J.R. 1985. Conservation of plants within their native ecosystems, pp. 197–219. *In* C. Gómez-Campo (ed.), *Plant Conservation in the Mediterranean.* Junk, The Hague, The Netherlands.

Tomaselli, R. 1977. Degradation of the Mediterranean maquis, pp. 33–72. *Mediterranean Forests and Maquis: Ecology, Conservation and Management.* MAB Technical, Note 2, UNESCO, Paris, France.

Trabaud, L. 1981. Man and fire: impacts of Mediterranean vegetation, pp. 523–537. *In* F. Di Castri, D.W. Goodall, and R.L. Specht (eds.), *Ecosystems of the World 11: Mediterranean Type Shrublands.* Elsevier, Amsterdam, The Netherlands.

Trabaud, L. 1983. The effects of different fire regimes on soil nutrient levels in *Quercus coccifera* garrigue, pp. 233–243. *In* F.J. Kruger, D.T. Mitchell, and J.U.M. Jarvis (eds.), *Mediterranean-Type Ecosystems: The Role of Nutrients.* Springer-Verlag, Berlin, Germany.

Vogl, R.J. 1980. The ecological factors that produce perturbation-dependent ecosystems, pp. 233–243. *In* J. Cairns, Jr. (ed.), *The Recovery Process in Damaged Ecosystems.* Science Publishers Inc., Ann Arbor, MI.

Voss, W., Stortelder, A. 1992. Vanishing Tuscan Landscapes. Pudoc Scientific Publishers, Wageningen, The Netherlands.

Waddington, C.H. 1977. *Tool for Thought.* Paladin. Frogmore, England.

Walter, H. 1968. Die Vegetation der Erde in Oekophysiologischer Betrachtung, Vol. 2. Fischer, Jena.

Walter, H. 1977. Effects of fire on wildlife communities. *Proceedings of the Symposium on Environmental Consequences of Fire and Fuel Management in Mediterranean Ecosystems, August 1977, Palo Alto, CA.* USDA For. Serv. Gen. Tech. Rep. WQO–3, Washington, DC.

Zohar, Y., Midani, E., Kutiel, P., Tsraeli, A. 1990. Prescribed burning as a tool in forest management. *Environmental Quality and Ecosystem Stability, Proc. Fifth. Intern. Conf. Israel Society for Ecology* and *Environmental Quality Sciences.*

Zohary, M. 1962. Plant Life in Palestine – Israel and Jordan. The Ronald Press Company, New York.

Index

Ecological Studies

Ecological Studies